T0146295

Design of Liquid Retaining Concrete Structures

Design of Liquid Retaining Concrete Structures

Third Edition

J.P. Forth BEng (Hons), PhD, CEng, MIStructE
Senior Lecturer in Structures, School of Civil Engineering,
University of Leeds
and
A.J. Martin BEng (Hons), MSt, CEng, MICE, MIStructE
Chartered Civil and Structural Engineer

Whittles Publishing

Published by
Whittles Publishing,
Dunbeath,
Caithness KW6 6EG,
Scotland, UK
www.whittlespublishing.com

First published in Great Britain 1981; Second edition 1992
ISBN 978-184995-052-7

Printed and bound in Malta by Melita Press

Contents

Preface

In 2010, a new suite of design codes was introduced into the UK. As such, the British Standard Codes of Practice 8110 *Structural Use of Concrete* and 8007 *Design of Concrete Structures for Retaining Aqueous Liquids* were replaced by Eurocode 2 (BS EN 1992-1-1) and Eurocode 2 Part 3 (BS EN 1992–3), respectively, both with accompanying UK specific National Application Documents. The guidance provided by these new codes is quoted as being much more theoretical in its nature and is therefore fundamentally different to the traditional step-by-step guidance that has been offered for many years in the UK by the British Standards. The approach of these new replacement codes is therefore a step change in design guidance, requiring much more interpretation.

The third edition of this book, whilst adopting a similar structure to the first two editions, has attempted to reflect this more theoretical approach. The new codes represented an opportunity to improve the guidance, based on a greater depth of research and practical experience gained over the last two decades. Unfortunately, the improvements are not as extensive as would have been hoped, partly because much research to corroborate some of the proposed new theory is still ongoing. In order to accommodate this position, the book offers an insight into some of the remaining shortcomings of the code and the potential improvements to the efficiency of design and possible innovations that are possible and which can hopefully be included in the planned revision of the codes in 2020.

JPF and AJM

Acknowledgements

I met Andrew Beeby for the first time in 1997; later, in 1999 the opportunity arose for me to join the Structures Group at the University of Leeds; I took up the position because Andrew was the head of that group. I have always felt privileged to have been able to call Andrew my mentor, a role which continued even after he retired; at which point in time I could more accurately and proudly call him my friend. I have never known anyone more insightful. His passing in 2011 was an extremely sad time. He was a true gentleman, possessing rare qualities; I give my thanks for his guidance, knowledge, motivation and friendship.

I would also like to thank all the engineers and researchers who have contributed to the better understanding of this fascinating topic of water retaining structures, past and present.

JPF, Leeds

Structural engineering is a fascinating subject and I acknowledge with grateful thanks all those who have influenced my education, training and development as an engineer throughout my career. I am grateful to Matt Kirby for permission to use the photograph reproduced in Figure 1.2. My contribution to this book is dedicated to my family and especially to my father Geoffrey H. Martin (1929–2013).

AJM, Copenhagen

We are both very grateful to Bob Anchor for this opportunity to produce the third edition of his book. His contribution to the design of water retaining structures is now into its fifth decade – an outstanding achievement.

Chapter 1
Introduction

1.1 Scope

It is common practice to use reinforced or prestressed concrete structures for the storage of water and other aqueous liquids. Similar design methods may also be used to design basements in buildings where groundwater must be excluded. For such purposes as these, concrete is generally the most economical material of construction and, when correctly designed and constructed, will provide long life and low maintenance costs. The design methods given in this book are appropriate for the following types of structure (all of which are in-line with the scope of Part 3 of Eurocode 2, BS EN 1992-3, 2006): storage tanks, reservoirs, swimming pools, elevated tanks (not the tower supporting the tank), ponds, settlement tanks, basement walls, and similar structures (Figures 1.1 and 1.2). Specifically excluded are: dams, structures subjected to dynamic forces, and pipelines, aqueducts or other types of structure for the conveyance of liquids.

It is convenient to discuss designs for the retention of water, but the principles apply equally to the retention of other aqueous liquids. In particular, sewage tanks are included. The pressures on a structure may have to be calculated using a specific gravity greater than unity, where the stored liquid is of greater density than water. Throughout this book it is assumed that water is the retained liquid unless any other qualification is made. The term 'structure' is used in the book to describe the vessel or container that retains or excludes the liquid.

The design of structures to retain oil, petrol and other penetrating liquids is not included (the code (BS EN 1992-3, 2006) recommends reference to specialist literature) but the principles may still apply. Likewise, the design of tanks to contain hot liquids (> 200°C) is not discussed.

1.2 General design objectives

A structure that is designed to retain liquids must fulfil the requirements for normal structures in having adequate strength, durability, and freedom from excessive cracking or deflection. In addition, it must be designed so that the liquid is not allowed to leak or percolate through the concrete structure. In the design of normal building structures, the most critical aspect of the design is to ensure that the structure retains its stability under the applied (permanent and variable) actions. In the design of structures to retain liquids, it is usual to find that if the structure has been proportioned and reinforced so that the liquid is retained without leakage (i.e. satisfying the Serviceability Limit State, SLS), then the strength (the Ultimate Limit State, ULS requirements)

Figure 1.1 *A tank under construction* (Photo: J.P. Forth/A.P. Lowe).

Figure 1.2 *A concrete tank (before construction of the roof) illustrating the simplicity of the structural form* (Photo: M.J. Kirby).

is more than adequate. The requirements for ensuring a reasonable service life for the structure without undue maintenance are more onerous for liquid-retaining structures than for normal structures, and adequate concrete cover to the reinforcement is essential. Equally, the concrete itself must be of good quality, and be properly compacted: good workmanship during construction is critical.

Potable water from moorland areas may contain free carbon dioxide or dissolved salts from the gathering grounds, which attack normal concrete. Similar difficulties may occur with tanks that are used to store sewage or industrial liquids. After investigating by tests the types of aggressive elements that are present, it may be necessary to increase the cover, the cement content of the concrete mix, use special cements or, under 'very severe' (BS EN 1992-1-1, 2004; BS 8500-1, 2006) conditions, use a special lining to the concrete tank.

1.3 Fundamental design methods

Historically, the design of structural concrete was based on elastic theory, with specified maximum design stresses in the materials at working loads. In the 1980s, limit state philosophy was introduced in the UK, providing a more logical basis for determining factors of safety. 2011 has seen the introduction of the new Eurocodes; BS 8110 and BS 8007 have been withdrawn, and in their place is a suite of new codes, including specifically BS EN 1992-1-1:2004 (Eurocode 2 Part 1 or EC2) and BS EN 1992-3: 2006 (Eurocode 2 Part 3 or EC2 Part 3) and their respective National Annexes. The new Eurocodes continue to adopt the limit state design approach. In ultimate design, the working or characteristic actions are enhanced by being multiplied by *partial safety factors*. The enhanced or ultimate actions are then used with the failure strengths of the materials, which are themselves modified by their own partial factors of safety, to design the structure.

Limit state design methods enable the possible modes of failure of a structure to be identified and investigated so that a particular premature form of failure may be prevented. Limit states may be 'ultimate' (where ultimate actions are used) or 'serviceability' (where service actions are used).

Previously, when the design of liquid-retaining structures was based on the use of elastic design (BS 5337), the material stresses were so low that no flexural tensile cracks developed. This led to the use of thick concrete sections with copious quantities of mild steel reinforcement. The probability of shrinkage and thermal cracking was not dealt with on a satisfactory basis, and nominal quantities of reinforcement were specified in most codes of practice. It was possible to align the design guidance relating to liquid-retaining structures with that of the Limit State code BS 8110 Structural Use of Concrete once analytical procedures had been developed to enable flexural crack widths to be estimated and compared with specified maxima (Base *et al.*, 1966; Beeby, 1979) and a method of calculating the effects of thermal and shrinkage strains had been published (Hughes, 1976).

Prior to the introduction of BS 8007 in the 1980s, BS 5337 allowed designers to choose between either elastic or limit state design. It has often been said 'A structure does not know how it has been designed'. Any design system that enables a serviceable structure to be constructed safely and with due economy is acceptable. However, since BS 8007 was introduced in the UK, limit state design has been used consistently and perhaps more successfully for the design of liquid-retaining structures and, although it has now been withdrawn, there is no reason why this trend cannot continue with the introduction of these new Eurocodes, which continue to utilise this limit state design philosophy.

1.4 Codes of practice

Guidance for the design of water-retaining structures can be found in BS EN 1992-3 which provides additional guidance, specific to containment structures, to that found in BS EN 1992-1-1 (BS EN 1992-3 does not provide guidance on joint detail). This approach is not unusual as the superseded code BS 8007 also provided additional rules to those found in the over-arching Structural Use of Concrete code, BS 8110. However, whereas BS 8110 contained both guidance on the philosophy of design and the loads and their combinations to be considered in design, a different approach is adopted in the Eurocodes. BS EN 1992-1-1 is itself supported by the Eurocode (BS EN 1990:2002–commonly referred to as Eurocode 0) Basis of Structural Design and Eurocode 1(BS EN 1991–10 parts) Actions on Structures. BS EN 1990 guides the designer in areas of structural safety, serviceability and durability–it relates to all construction materials. BS EN 1991 actually supersedes BS 6399 Loading for Buildings and BS 648 Schedule of weights of building materials. All Eurocodes and their individual Parts are accompanied by a National Annex (NA) / National Application Document (NAD), which provide guidance specific to each individual state of the European Union, i.e. the UK National Application Document only applies to the UK. Values in these National Annexes may be different to the main body of text produced in the Eurocodes by the European Committee for Standardization (CEN).

There are two distinct differences between BS 8110/BS 8007 and the new Eurocodes, which will immediately be apparent to the designer. Eurocodes provide advice on structural behaviour (i.e. bending, shear etc.) and not member types (i.e. beams etc.). Also, Eurocodes are technically strong and fundamental in their approach–they do not provide a step-by-step approach on how to design a structural member.

1.5 Impermeability

Concrete for liquid-retaining structures must have low permeability. This is necessary to prevent leakage through the concrete and also to provide adequate durability, resistance to frost damage, and protection against corrosion for the reinforcement and other embedded steel. An uncracked concrete slab of adequate thickness will be impervious to the flow of liquid if the concrete mix has been properly designed and compacted into position. The specification of suitable concrete mixes is discussed in Chapter 2. Practically, the minimum thickness of poured in-situ concrete for satisfactory performance in most structures is 300 mm. Thinner slabs should only be used for structural members of very limited dimensions or under very low liquid pressures.

Liquid loss may occur at joints that have been badly designed or constructed, and also at cracks or from concrete surfaces where incomplete compaction has been achieved. It is nearly inevitable that some cracking will be present in all but the simplest and smallest of structures. If a concrete slab cracks for any reason, there is a possibility that liquid may leak or that a wet patch will occur on the surface. However, it is found that cracks of limited width do not allow liquid to leak (Sadgrove, 1974) and the problem for the designer is to limit the surface crack widths to a predetermined size. Cracks due to shrinkage and thermal movement tend to be of uniform thickness (although this does depend on the uniformity of the internal restraint) through the thickness of the slab, whereas cracks due to flexural action are of limited depth and are backed up by a depth of concrete that is in compression. Clearly, the former type of crack is more serious in allowing leakage to occur.

An important question is whether or not the cracks formed from the two cases mentioned above (Early Thermal and Loading) are additive. It is accepted that long-term effects may be complementary to early thermal cracking and in these instances steps are taken to reduce the limiting crack width for early deformations. However, currently there is no suggestion or process by which cracking resulting from early-age effects should be added to that resulting from structural loading. It has to be said that no problems have been recognised specific to this; however, it does not mean that it is not occurring. In fact, recent investigations by the author into shrinkage curvature have suggested that both extension of early age cracks and new cracks can occur on loading (Forth *et al.*, 2004).

Before considering whether or not early-age cracking is additive with cracking from structural loading it is worth clarifying the conditions of external restraint to imposed deformation, which can result in this early-age cracking. This external restraint results from either end or edge (base) restraint. Figure 1.3 illustrates the two forms of restraint. These two types of restraint are really limiting forms of restraint. In practice, the situation is somewhat more complicated and the actual restraint is either a combination of these two forms or, more likely when early thermal movements are being considered in a wall, one of edge restraint (Beeby and Forth, 2005).

An example of where both forms of restraint exist can be found by considering a new section of concrete cast between two pre-existing concrete wall sections and onto a pre-existing concrete base. At the base, edge restraint will dominate (see Figure 1.4–Zone 2). However, further up the wall away from the base, edge restraint will become less significant and end restraint will become more influential. At a point within the height of the wall, end restraint will dominate and edge restraint becomes insignificant (see Figure 1.4–Zone 1). The position and significance of the two restraint conditions

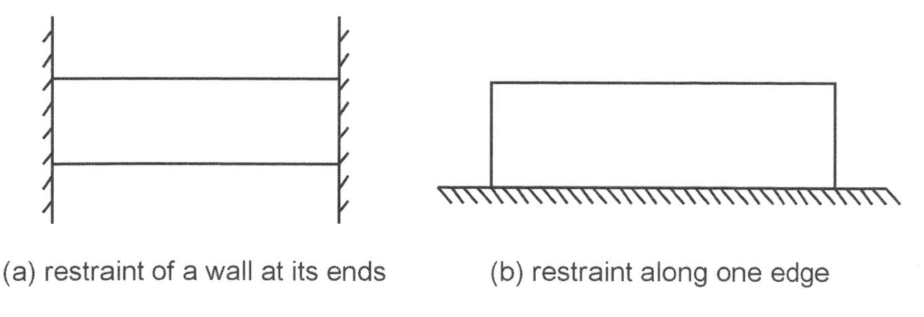

Figure 1.3 *External end and edge (base) restraint.*

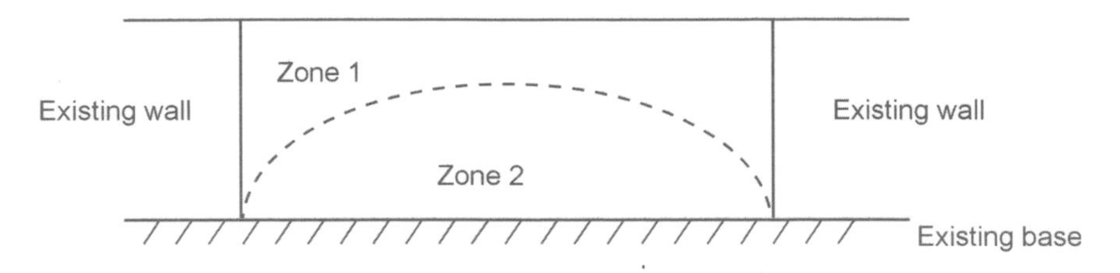

Figure 1.4 *Approximate regions of domination of end (Zone 1) and edge (Zone 2) restraint in an infill wall.*

is obviously dependent on the height, cross section and length of the concrete section as well as the concrete base.

BS EN 1992-3 provides restraint factors, R for various wall and floor slab placing sequences (this figure is reproduced from BS 8007). Diagrammatically it attempts to describe the combination of the two types of external restraint described above, i.e. end and edge restraint, although the restraint factor, R is really only based on the structural model of a member restrained at its end against overall shortening.

On the matter of whether or not early age cracking can be compounded by load cracking, consider the example of a horizontal slab between rigid end restraints (Fig. L1 (b) of BS EN 1992-3). Due to end restraint conditions, a slab between rigid restraints will produce a primary crack, parallel to the rigid restraints most likely midway between the restraints. This is also the most likely position of a crack to form from structural loading. So although further investigations are required to confirm the presence of combined cracking, clearly in this case the opportunity exists.

In the case of a wall cast on a base, if the wall is sufficiently long then even without the restraint offered by adjacent wall panels a primary vertical crack may develop due to the edge restraint of early age movement. Structurally the wall will behave as a cantilever and structural cracking will therefore be horizontal in nature. In such a case, it is clear that early age cracking is not compounded by structural cracking. Taking this example one step further and considering Fig. L1 (d) of BS EN 1992-3, which illustrates a wall restrained at its base and by adjacent wall panels, diagonal cracks are predicted to occur at the base of the wall and near its ends. It is unsure as to whether these diagonal cracks would influence the formation and behaviour of structural cracking; further investigation is required.

As mentioned above, no problems have been identified that can be specifically explained by this potential combination of early-age and structural cracking. This could be because fortuitously, the code guidance for the design of water-retaining structures results in an over-estimation of steel required to resist imposed deformations. For edge-restrained situations, the crack width depends on the restrained imposed strain and not the tensile strength of the concrete (Al Rawi and Kheder, 1990). The amount of horizontal reinforcement is entirely dictated by that needed to control early thermal cracking (restraint to early thermal movement). Traditional detailing used about 0.2% of anti-crack reinforcement, whereas BS 8007 tended to require at least twice this amount (because of the intended use of the structure and the better control of crack widths required in water-retaining structures). The Eurocodes appear to require between 0.3 and 0.4%. These all relate to restraint of early thermal movement which, as discussed earlier, is based on the end restraint condition and not edge restraint. The question is one of whether this amount of steel is actually necessary.

1.6 Site conditions

The choice of site for a reservoir or tank is usually dictated by requirements outside the structural designer's responsibility, but the soil conditions may radically affect the design. A well-drained site with underlying soils having a uniform safe bearing pressure at foundation level is ideal. These conditions may be achieved for a service reservoir near to the top of a hill, but at many sites where sewage tanks are being constructed, the subsoil has a poor bearing capacity and the groundwater table is near to the surface. A high level of groundwater must be considered in designing the tanks in order to prevent flotation (Figure 1.5), and poor bearing capacity may give rise to increased settlement. Where the subsoil strata dip, so that a level excavation intersects more than one type of subsoil, the effects of differential settlement must be considered (Figure 1.6). A soil survey is always necessary unless an accurate record of the subsoil is available. Typically, boreholes of at least 150 mm diameter should be drilled to a depth of 10 m, and soil samples taken and tested to determine the sequence of strata and the allowable bearing pressure at various depths. The information from boreholes should be supplemented by digging trial pits with a small excavator to a depth of 3–4 m.

The soil investigation must also include chemical tests on the soils and groundwater to detect the presence of sulphates or other chemicals in the ground that could attack the concrete and eventually cause corrosion of the reinforcement (Newman and Choo, 2003). Careful analysis of the subsoil is particularly important when the site has previously been used for industrial purposes, or where groundwater from an adjacent tip may flow through the site. Further information is given in Chapter 2.

When mining activity is suspected, a further survey may be necessary and a report from the mineral valuer or a mining consultant is necessary. Deeper, randomly located boreholes may be required to detect any voids underlying the site. The design of a reservoir to accept ground movement due to future mining activity requires the provision of extra movement joints or other measures to deal with the anticipated movement and is outside the scope of this book (Davies, 1960; Melerski, 2000). In some parts of the world, consideration must be given to the effects of earthquakes, and local practice should be ascertained.

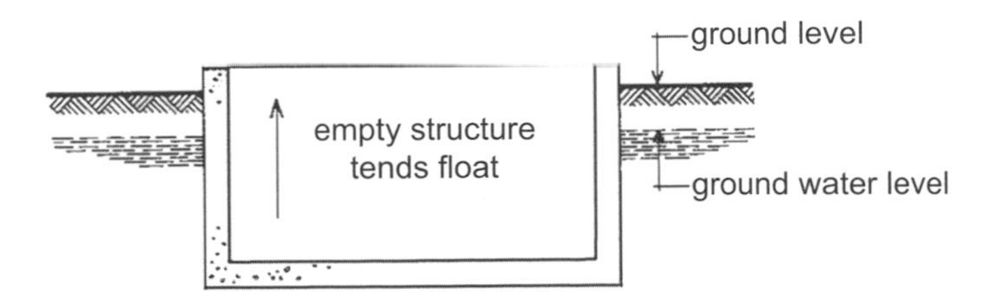

Figure 1.5 *Tank flotation due to ground water.*

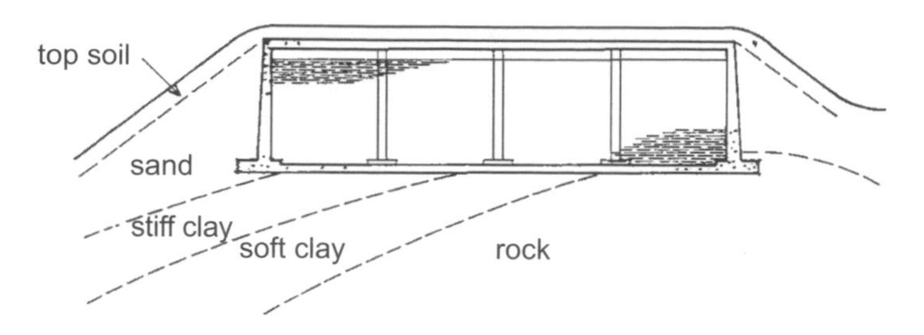

Figure 1.6 *Effect of varying strata on settlement.*

1.7 Influence of execution methods

Any structural design has to take account of the constructional problems involved and this is particularly the case in the field of liquid-retaining structures. Construction joints in building structures are not normally shown on detailed drawings but are described in the specification. For liquid-retaining structures, construction joints must be located on drawings, and the contractor is required to construct the works so that concrete is placed in one operation between the specified joint positions. The treatment of the joints must be specified, and any permanent movement joints must be fully detailed. All movement joints require a form of waterstop to be included; construction joints may or may not be designed using a waterstop (BS 8102:2009). Details of joint construction are given in Chapter 5. In the author's opinion, the detailed design and specification of joints is the responsibility of the designer and not the contractor. The quantity of distribution reinforcement in a slab and the spacing of joints are interdependent. Casting one section of concrete adjacent to another section, previously cast and hardened, causes restraining forces to be developed that tend to cause cracks in the newly placed concrete. It follows that the quantity of distribution reinforcement also depends on the degree of restraint provided by the adjacent panels.

Any tank that is to be constructed in water-bearing ground must be designed so that the groundwater can be excluded during construction. The two main methods of achieving this are by general ground de-watering, or by using sheet piling. If sheet piling is to be used, consideration must be given to the positions of any props that are necessary, and the sequence of construction that the designer envisages (Gray and Manning, 1973).

1.8 Design procedure

As with many structural design problems, once the member size and reinforcement have been defined, it is relatively simple to analyse the strength of a structural member and to calculate the crack widths under load: but the designer has to estimate the size of the members that he proposes to use before any calculations can proceed. With liquid-retaining structures, crack-width calculations control the thickness of the member, and therefore it is impossible to estimate the required thickness directly unless the limited stress method of design is used.

An intermediate method of design is also possible where the limit state of cracking is satisfied by limiting the reinforcement stress rather than by preparing a full calculation. This procedure is particularly useful for sections under combined flexural and direct stresses.

1.9 Code requirements (UK)

BS EN 1992-3 is based on the recommendations of BS EN 1992-1-1 for the design of normal structural concrete, and the design and detailing of liquid-retaining structures should comply with BS EN 1992-1-1 except where the recommendations of BS EN 1992-3 (and the UK National Annex) vary the requirements. The modifications that have been introduced into the Eurocodes mainly relate to:

- surface zones for thick sections with external restraint;
- surfacc zoncs for internal restraint only;
- the critical steel ratio, ρ_{crit};
- the maximum crack spacing, $S_{r,max}$;
- edge restraint.

These modifications are suitably discussed by Bamforth (2007), Hughes (2008) and Forth (2008).

Chapter 2
Basis of design and materials

2.1 Structural action

It is necessary to start a design by deciding on the type and layout of structure to be used. Tentative sizes must be allocated to each structural element, so that an analysis may be made and the sizes confirmed.

All liquid-retaining structures are required to resist horizontal forces due to the liquid pressures. Fundamentally there are two ways in which the pressures can be contained:

(i) by forces of direct tension or compression (Figure 2.1);
(ii) by flexural resistance (Figure 2.2).

Structures designed by using tensile or compressive forces are normally circular and may be prestressed (see Chapter 4). Rectangular tanks or reservoirs rely on flexural action using cantilever walls, propped cantilever walls or walls spanning in two directions. A structural element acting in flexure to resist liquid pressure reacts on the supporting elements and causes direct forces to occur. The simplest illustration (Figure 2.3) is a small tank. Additional reinforcement is necessary to resist such forces unless they can be resisted by friction on the soil.

2.2 Exposure classification

Structural concrete elements are exposed to varying types of environmental conditions. The roof of a pumphouse is waterproofed with asphalt or roofing felt and, apart from a short period during construction, is never externally exposed to wet or damp conditions. The exposed legs of a water tower are subjected to alternate wetting and drying from rainfall but do not have to contain liquid. The lower sections of the walls of a reservoir are always wet (except for brief periods during maintenance), but the upper sections may be alternately wet and dry as the water level varies. The underside of the roof of a closed reservoir is damp from condensation–because of the waterproofing on the external surface of the roof, the roof may remain saturated over its complete depth. These various conditions are illustrated in Figure 2.4.

Experience has shown that, as the exposure conditions become more severe, precautions should be taken to ensure that moisture and air do not cause carbonation in the concrete cover to the reinforcement thus removing the protection to the steel and causing corrosion, which in turn will cause the concrete surface to spall (Newman, 2003). Adequate durability can normally be ensured by providing a dense well-compacted concrete mix (see Section 2.5.2) with a concrete cover (cast against formwork) in the

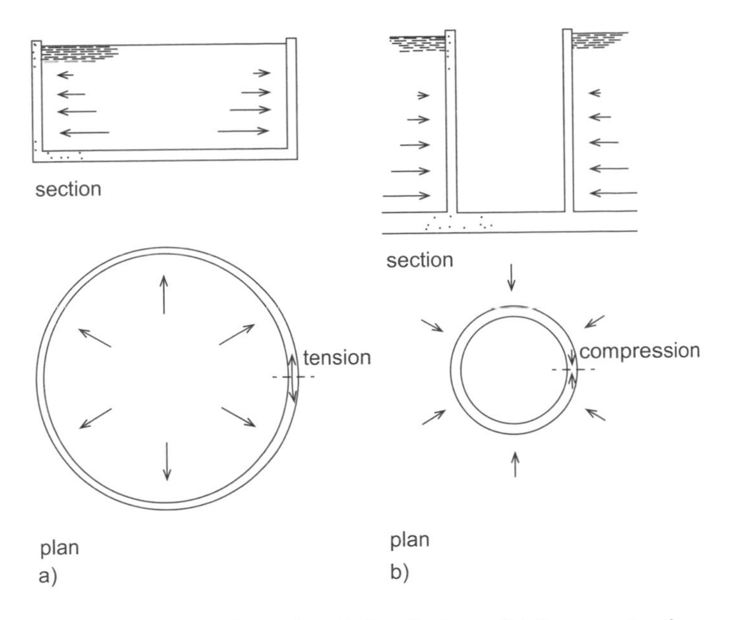

Figure 2.1 *Direct forces in circular tanks. (a) Tensile forces (b) Compressive forces.*

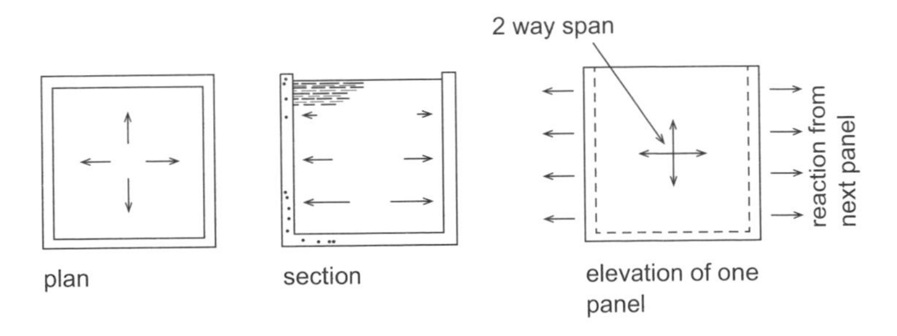

Figure 2.2 *Direct forces of tension in wall panels of rectangular tanks.*

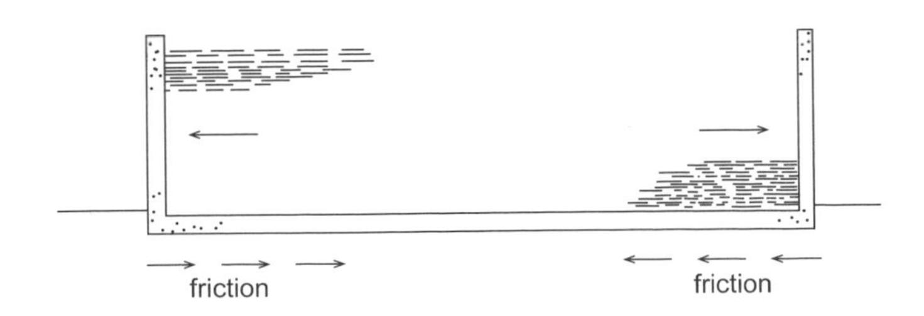

Figure 2.3 *Tension in floor of a long tank with cantilever walls.*

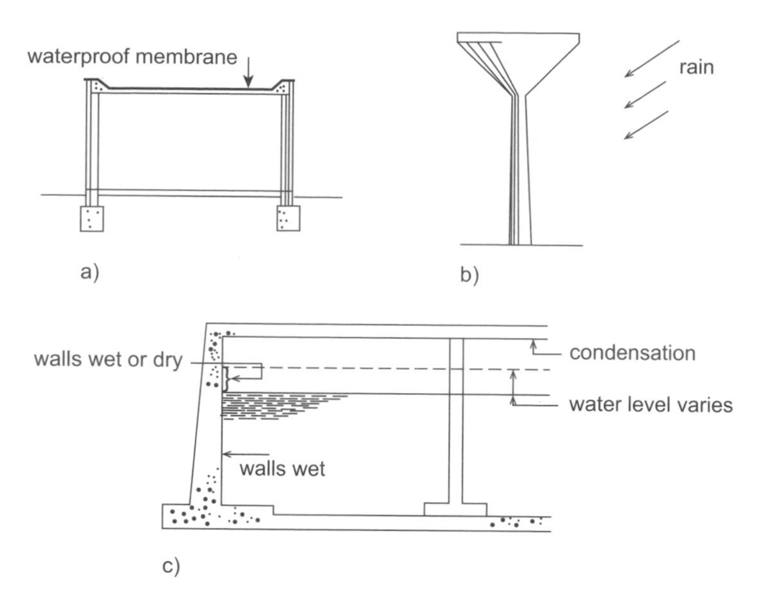

Figure 2.4 *Exposure to environmental conditions: (a) pumphouse roof, (b) water tower and (c) reservoir.*

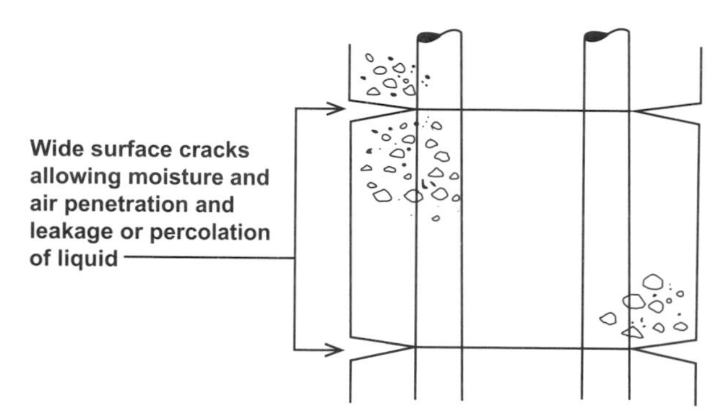

Figure 2.5 *Effect of cracks.*

region of at least 40 mm (BS 8500-1), but it is also necessary to control cracking in the concrete, and prevent percolation of liquid through the member (see Figure 2.5).

Previously, for design purposes, BS 8110 conveniently classified exposure in terms of relative severity (i.e. mild, moderate, severe). However, exposure classification in Eurocode 2 is now related to the deterioration processes, i.e. carbonation, ingress of chlorides, chemical attack from aggressive ground and freeze/thaw. Acting alongside Eurocode 2 is a more comprehensive guide, BS 8500 (Parts 1 and 2), to assist in determining cover. For less severe exposure conditions, BS 8500 is perhaps less onerous than BS 8110. However, for more severe conditions the requirements of BS 8500 are different. This is important, as BS EN 1992-3 requires that all liquid-retaining structures should be designed for at least 'severe' conditions of exposure. Where appropriate the 'very severe' and 'extreme' categories should be used. As an example, a water tower near to the sea coast and exposed to salt water spray would be designed for 'very severe' exposure.

As well as defining cover, durability requirements are also achieved by controlling cracking. For the serviceability limit state, the maximum (limiting) crack width is between 0.05 mm and 0.2 mm, depending on the ratio of the hydrostatic pressure to wall thickness. It should be noted that these limiting crack widths are actually equivalent to total crack width, i.e. in theory, early age, long term and loading (see comments in Chapter 1). The range of crack widths provided above is provided in BS EN 1992-3. General guidance on crack control is provided in Section 7.3 of BS EN 1992-1-1. Additional guidance is given in BS EN 1992-3 because of the nature of the structure. Early age thermal cracking may result in through cracks, which can lead to seepage or leakage. In water-retaining structures this could be deemed a failure. BS EN 1992-3 therefore provides a 'Classification of Tightness', shown below in Table 2.1. This tightness represents the degree of protection against leakage: 0 (zero) represents general provision for crack control in-line with BS EN 1992-1-1; 3 represents no leakage permitted. Tightness class 1 is normally acceptable for water-retaining structures.

The requirement for 'No leakage permitted' does not mean that the structure will not crack but simply that the section is designed so that there are no through cracks. There is no crack width recommendation of 0.1 mm for critical aesthetic appearance in the new Eurocodes as there was in BS 8110. No rational basis for defining the aesthetic appearance of cracking exists. BS EN 1992-3 claims that for Tightness class 1 structures, limiting the crack widths to the appropriate value within the range stated above should result in the effective sealing of the cracks within a relatively short time. The ratios actually represent pressure gradients across the structural section. As such, the claim that cracks of 0.2 mm will 'heal' provided that the pressure gradient does not exceed 5 has not changed much to the claim in BS 8007. For crack widths of less than 0.05 mm, healing will occur even when the pressure gradient is greater than 35. The fact that these cracks do seal is not strictly only due to autogenous healing (i.e. self-healing due to formation of hydration products) as was claimed in BS 8007, but also possibly due to the fact that the crack becomes blocked with fine particles. As mentioned above, sealing under hydrostatic pressure is discussed in Clause 7.3.1 of BS EN 1992-3 and for serviceability conditions, the limit state appropriate for water retaining structures, crack widths are limited to between 0.05 and 0.2 mm. When considering appearance and durability, further guidance with respect to crack widths and their relationship with exposure conditions can be found in Clause 7.3.1 of BS EN 1992-1-1 and its NA (Table NA.4).

Table 2.1 *Tightness classification.*

Tightness class	*Requirements for leakage*
0	Some degree of leakage acceptable, or leakage of liquids irrelevant.
1	Leakage to be limited to a small amount. Some surface staining or damp patches acceptable.
2	Leakage to be minimal. Appearance not to be impaired by staining.
3	No leakage permitted

2.3 Structural layout

The layout of the proposed structure and the estimation of member sizes must precede any detailed analysis. Structural schemes should be considered from the viewpoints of strength, serviceability, ease of construction, and cost. These factors are to some extent mutually contradictory, and a satisfactory scheme is a compromise, simple in concept and detail. In liquid-retaining structures, it is particularly necessary to avoid sudden changes in section, because they cause concentration of stress and hence increase the possibility of cracking.

It is a good principle to carry the structural loads as directly as possible to the foundations, using the fewest structural members. It is preferable to design cantilever walls as tapering slabs rather than as counterfort walls with slabs and beams. The floor of a water tower or the roof of a reservoir can be designed as a flat slab. Underground tanks and swimming-pool tanks are generally simple structures with constant-thickness walls and floors.

It is essential for the designer to consider the method of construction and to specify on the drawings the position of all construction and movement joints. This is necessary as the detailed design of the structural elements will depend on the degree of restraint offered by adjacent sections of the structure to the section being placed. Important considerations are the provision of 'kickers' (or short sections of upstand concrete) against which formwork may be tightened, and the size of wall and floor panels to be cast in one operation.

2.4 Influence of construction methods

Designers should consider the sequence of construction when arranging the layout and details of a proposed structure. At the excavation stage, and particularly on water-logged sites, it is desirable that the soil profile to receive the foundation and floors should be easily cut by machine. Flat surfaces and long strips are easy to form but individual small excavations are expensive to form. The soil at foundation level exerts a restraining force (the force develops from the restraint of early thermal contraction and shrinkage) on the structure, which tends to cause cracking (Figure 2.6). The

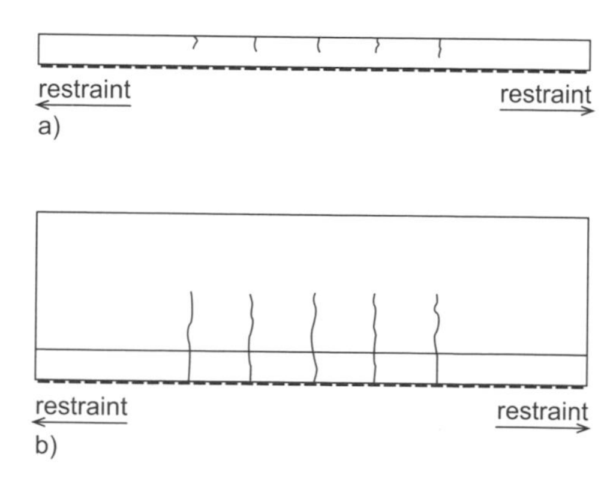

Figure 2.6 *Cracking due to restraint by frictional forces at foundation level (a) Floor slab (b) Wall (indicative only).*

frictional forces can be reduced by laying a sheet of 1 000 g polythene or other suitable material on a 75 mm layer of 'blinding' concrete. For the frictional forces to be reduced, it is necessary for the blinding concrete to have a smooth and level surface finish. This can only be achieved by a properly screeded finish, and in turn this implies the use of a grade of concrete that can be so finished (BS 8500-1, 2006; Teychenne, 1975; Palmer, 1977). A convenient method is to specify the same grade of concrete for the blinding layer as is used for the structure. This enables a good finish to be obtained for the blinding layer, and also provides an opportunity to check the strength and consistency of the concrete at a non-critical stage of the job. It also reduces the nominal cover, c_{nom} (BS 8500-1, 2006).

The foundations and floor slabs are constructed in sections that are of a convenient size and volume to enable construction to be finished in the time available. Sections terminate at a construction or movement joint (Chapter 5). The construction sequence should be continuous as shown in Figure 2.7(a) and not as shown in Figure 2.7(b). By adopting the first system, each section that is cast has one free end and is enabled to shrink on cooling without end restraint (a day or two after casting), although edge restraint will still exist (see Chapters 1 and 5). With the second method, considerable tensions are developed between the relatively rigid adjoining slabs.

Previously, BS 8007 provided three design options for the control of thermal contraction and restrained shrinkage: continuous (full restraint), semi-continuous (partial restraint) and total freedom of movement. On the face of it, it appears that BS EN 1992-3 does not allow semi-continuous design and therefore partial contraction joints have been excluded. Therefore, Part 3 only offers two options: full restraint (no movement joints) and free movement (minimum restraint). For the condition of free movement, Part 3 recommends that complete joints (free contraction joints) are spaced at the greater of 5 m or 1.5 times the wall height. (This is similar to the maximum crack spacing of a wall, given in BS EN 1992-1-1 Section 7, with no or less than $A_{s,\,min}$ bonded reinforcement within the tension zone, i.e. 1.3 times the height of the wall.) However, BS EN 1992-3 also states 'a moderate amount of reinforcement is provided sufficient to transmit any movements to the adjacent joint'. This appears contradictory. Hence continuity steel, less than $A_{s,\,min}$ is still permitted and semi-continuous joints are therefore still allowed.

Figure 2.7 *Construction sequence (a) Preferred sequence (b) Not recommended (c) Effect of method (b) on third slab panel (cracks shown are illustrative only).*

It is recommended that if partial contraction joints are used, continuity steel of at least $A_{s, min}$ (or 50% of the full continuity steel) is used and in fact the recommended spacing of these partial contraction joints is similar to that proposed previously (approx 7.5 m). BS EN 1992-3 does not actually provide any guidance on the spacing of full contraction joints, i.e. no continuity steel (the 15m between full contaction joints shown in part (a) of Figure 2.8 below illustrates the guidance previously available in BS 8007).

Alternatively, temporary short gaps may be left out, to be filled in after the concrete has hardened. A further possibility is the use of induced contraction joints, where the concrete section is deliberately reduced in order to cause cracks to form at preferred positions. These possibilities are illustrated in Figure 2.8. The casting sequence in the vertical direction is usually obvious. The foundations or floors are laid with a short section of wall to act as a key for the formwork (the kicker, Figure 2.9). Walls may be concreted in one operation up to about 8 m height.

Reinforcement should be detailed to enable construction to proceed with a convenient length of bar projecting from the sections of concrete, which are placed at each stage of construction. Bars should have a maximum spacing of 300 mm or the thickness of the slab and a minimum spacing dependent on size, but not usually less than 100 mm to allow easy placing of the concrete. Distribution or shrinkage

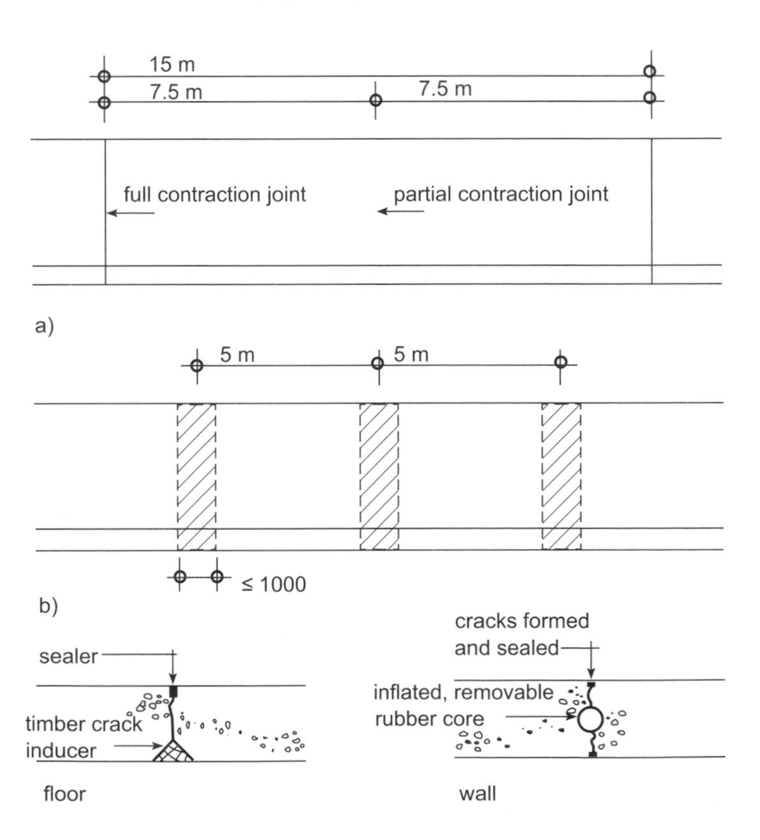

Figure 2.8 *Joints (a) Typical layout in a wall (b) Typical layout of temporary gaps in construction (c) Induced joints.*

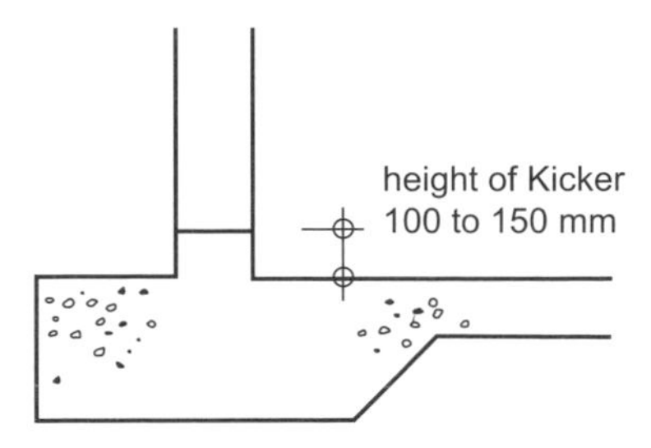

Figure 2.9 *Joint between floor and wall.*

reinforcement should ideally be placed in the outer layers nearest to the surface of the concrete. In this position it has maximum effect. However, structural considerations (i.e. to maximise effective depth, *d*) may also influence the layering of the reinforcement in each face. Figure 2.8 also illustrates typical crack inducers that can be used at full or partial contraction joints.

2.5 Materials and concrete mixes

2.5.1 Reinforcement

Although the service tensile stress in the reinforcement in liquid-retaining structures is not always very high, it is standard practice to specify high-strength steel with a ribbed or deformed surface either in single bar form or as mesh.

BS EN 1992-1-1 Annex C permits a range of characteristic yield strengths between 400 and 600 MPa; the specified characteristic strength of reinforcement available in the UK is 500 MPa. The specified characteristic strength is a statistical measure of the yield or proof stress of a type of reinforcement. The proportion of bars that fall below the characteristic strength level is defined as 5% (Figure 2.10). A material partial safety factor (for Persistent and Transient loading, $\gamma_m = 1.15$) is applied to the specified characteristic strength to obtain the ultimate design strength. In the UK, high-yield bars are supplied in accordance with BS 4449: 2005 and BS 8666:2005. Both of these codes support BS EN 10080: 2007 but are stand-alone documents with no confliction with EN 10080. Three grades of high-yield steel (A to C) are listed in BS 4449. These gradings reflect the ductility of the steel, with grade C being the most ductile (suitable for seismic applications). In the UK, B500 steel denotes high-yield steel with a characteristic strength of 500 MPa; B500B denotes Normal grade B steel. However, it is not unknown for steel manufacturers to supply Grade C quality steel under the heading of Normal grade B steel. Grade B500A steel is provided for cold working.

The fact that plain round grade 250 MPa steel was excluded from BS 4449 reflects the fact that other standards are available for the specification of mild steel bars (BS EN 10025-1, 2004; BS EN 13877-3, 2004) and the fact that this grade was being used less frequently (CARES, 2012).

Welded fabric or mesh reinforcement is specified in BS 4483 (2005). It is available in four types from A to D. For water-retaining structures, type A or Square Mesh is most common. Square Mesh is manufactured using 10, 8, 7 or 6 mm diameter bars at 200 mm centres in both the longitudinal and transverse directions.

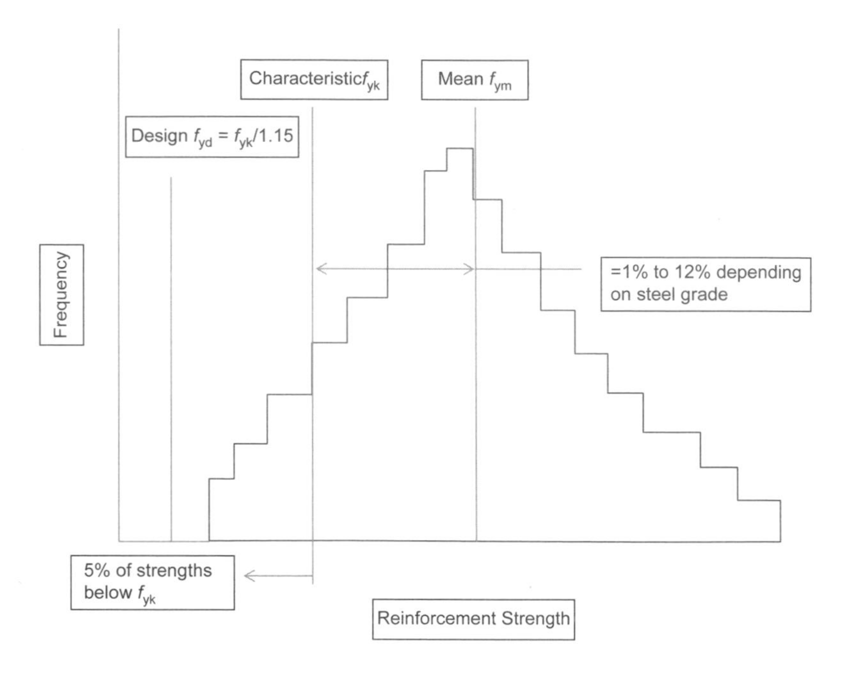

Figure 2.10 *Graphical definition of characteristic strength.*

Reinforcement embedded in concrete is protected from corrosion by the alkalinity of the cement. As time passes, the surface of the concrete reacts with carbon dioxide from the air and carbonates are formed that remove the protection. In certain circumstances, where perhaps restricted for space or where greater risk mitigation is required, special types of reinforcement can be considered. Stainless steel bars are still popular in the UK and are specified in BS 6722 (1986). The cost of stainless steel bars is still approximately 10 to 12 times that of normal grade high-yield bars. Hot dipped galvanising is still used to protect steel in some applications. In general, normal grade steel is fabricated into reinforcement cages first before being dipped. Due to the problems encountered with epoxy-coated bars, the use of this type of treatment to bars in the UK is almost non-existent.

2.5.2 Concrete

The specification, performance, production and conformity of concrete are controlled by BS EN 206-1, which was introduced in 2000 with subsequent amendments in 2004 and 2005. However, the UK NA to BS EN 1992-1-1 requires the use of BS 8500, which is a complementary British Standard to BS EN 206-1 and which contains additional UK provisions.

British Standard 8500 uses 'compressive strength classes' to define concrete strengths. Its notation uses both cylinder and cube strength (i.e. C25/30–cylinder / cube). It provides guidance on specifying concrete (cement type, aggregates, admixtures etc.) and an assessment of cover and strength for durability. As such, it replaces BS 5328 and related sections of BS 8110-1.

The detailed specification and design of concrete mixes is outside the scope of this book. However, a typical mix design for water-retaining structures is derived and provided in Chapter 6.

Cements

In recent years, there has been a great deal of progress made in improving the sustainability of concrete. This has been achieved through the manufacturing process of cements, the choice of aggregates and the reduction in CEM 1 (Portland cement) in the cements themselves. This reduction in volume of CEM 1 is achieved through the replacement of the Portland cement or by blending the Portland cement with other materials. However, the reduction in the overall volume of Portland cement is also driven by the desire to achieve concretes with greater variations in properties and performances (reduction in heat of hydration, quicker strength gain, greater frost resistance, reduced water content etc.). It should be noted that often an improvement in one property will be to the detriment of another–the designer can therefore have a lot to consider when considering both the technical and economic advantages that this new range of cements can provide. For instance, if the designer is required to specify a lower strength requirement it will not necessarily mean that the concrete will exhibit a loss of durability. Concretes containing cement replacement materials such as fly ash or ground granulated blastfurnace slag (GGBS) may offer better protection to rebar than CEM 1 concretes. An extensive investigation has been performed by Dhir *et al.* (2004); more information can also be obtained from Bamforth (2007).

Aggregates

The maximum size of aggregate must be chosen in relation to the thickness of the structural member. A maximum size of 20 mm is always specified up to member thickness of about 300–400 mm and may be used above this limit, particularly if larger aggregate sizes are not available. Size 40 may be specified in very thick members, if available. The use of a large maximum size of aggregate has the effect of reducing the cement content in the mix for a given workability, and hence reduces the amount of shrinkage cracking.

It is important to choose aggregates that have low drying shrinkage (the maximum drying shrinkage specified in BS 8500: Part 2 should be < 0.075%, unless otherwise specified) and low absorption. Most quartz aggregates are satisfactory in these respects but, where limestone aggregate is proposed, some check on the porosity is desirable. Certain aggregates obtained from igneous rocks exhibit high shrinkage properties and are quite unsuitable for use in liquid-retaining structures.

Thc aggregate type also has on influence on early thermal cracking in concrete. A preferred normal weight coarse aggregate is a crushed rock (crushed aggregates produce concretes with higher tensile strain capacity than rounded aggregates) with a low coefficient of thermal expansion. Typically, this applies to many limestones. Again, the designer has to consider carefully these factors when balancing the advantages and disadvantages. Even lower coefficients of thermal expansion are exhibited by lightweight aggregates; these produce concretes with even higher tensile strain capacities and are becoming more popular, as are the recycled concrete aggregates (RCA) and the recycled aggregates (RA), which are again specified in BS 8500 Parts 1 and 2.

Local suppliers can often provide evidence of previous use that will satisfy the specifier (in some instances, this is a requirement of BS 8500). Aggregates are expensive to transport and locally available material is preferable in terms of both cost and sustainability; however, some care is necessary when using material from a new quarry, and tests of the aggregate properties are recommended.

Admixtures

Admixtures are generally included to improve the strength and durability of the concrete. Typical admixtures are plasticisers and super-plasticisers, which can be used to increase the workability of the concrete, allowing it to be placed more easily with less consolidating effort, or to reduce the water content while maintaining workability (hence they are also known as water-reducing agents). An air-entraining agent is another admixture. These admixtures entrain air bubbles within the concrete, improving the freeze-thaw resistance of the concrete and hence its durability. However, whilst durability is improved, the strength may be reduced. As a general guide, for each 1% of entrained air there is a possible 5% reduction in compressive strength. Admixtures can also be added to slow the hydration of the cement, such as in large pours where partial setting before the pour is complete is clearly undesirable. Admixtures containing calcium chloride are not desirable as there is a risk of corrosion of the reinforcement.

Concrete mix design

The stages in the design of a concrete mix are as follows. Initially, the relevant exposure condition should be identified–each face of the structure and its individual element should be considered and apportioned an exposure class. Exposure classes in BS 8500 are related to the deterioration processes of carbonation (XC classes), freeze / thaw (XF classes), chloride ingress (XD and XS classes) and chemical attack, including sulphate attack, from aggressive ground. (BS 8500 refers the designer to the BRE Special Digest 1 (2005), which gives guidance on the assessment of the aggressive chemical environment for concrete class (ACEC), rather than the XA classes used in BS EN 206-1.) All of these X classes are sub-divided; it is likely that there will always be at least one relevant exposure class for each element.

Once the relevant exposure condition(s) have been identified, a strength class and cover (including permitted deviations) are chosen that will ensure a minimum 50-year working life of the structure.

The concrete must be designed to provide a mix that is capable of being fully compacted by the means available. Any areas of concrete that have not been properly compacted are likely to leak. The use of poker-type internal vibrators is recommended.

2.6 Loading

2.6.1 Actions

Characteristic values for actions (loads) are given in BS EN 1991 (Eurocode 1: Actions on Structures). Typically, liquid-retaining structures are subject to loading by pressure from the retained liquid. The nominal densities of materials are provided in BS EN 1992-1-1, however, this part does not provide the densities of all of the materials that may be stored in liquid-retaining structures. Table 2.2 provides the nominal density for typically retained liquids.

Table 2.2 *Nominal density of retained liquids.*

Liquid	*Weight (kN/m^3)*
Water	10.0
Raw sewage	11.0
Digested sludge aerobic	10.4
Digested sludge anaerobic	11.3
Sludge from vacuum filters	12.0

BS EN 1991 also provides specific guidance for Silos and Tanks (BS EN 1991-4). Guidance on Thermal actions (BS EN 1991-1-5) and Execution actions (BS EN 1991-1-6) can also be particularly relevant to the design of water-retaining structures.

External reservoir walls are also often required to support soil fill. The soil loading conditions to be considered are illustrated in Figure 2.11–actual soil loading depends on the water table condition, the state of compaction of the backfill and whether native soil is used for the backfill. In the long-term it is likely that pressures will approach the 'at-rest' situation, although clay backfill may take many years to mobilise. For design, when the reservoir is empty, full allowance must be made for the 'at rest' or active earth pressure with the appropriate partial safety factors, assuming the backfill is carefully controlled, and any surcharge pressures from vehicles. When designing for the 'reservoir full' case, as a minimum the active earth pressure should be presumed. It is important to note that when designing for the condition with the reservoir full, no relief should be allowed from passive pressure of the soil fill. This is because of the differing moduli of elasticity of soil and concrete, which prevent the passive resistance of the soil being developed before the concrete is fully loaded by the pressure from the contained liquid (effectively, not enough strain can be generated in the soil to produce the passive pressure; however, it does depend on the method of backfill utilised and if the soil is over-compacted it is possible to create a situation where pseudo passive conditions exist).

At ultimate limit state–persistent and transient situations–three separate sets of load combinations (i.e. combinations of permanent and variable actions) are provided by Eurocode 0. These are (i) EQU, to be used if the structure is to be checked against loss of equilibrium; (ii) STR, to be used to check internal failure of the structure as governed by the strength of the construction materials (note, for this combination, the strength can be considered when the design does and does not also involve geotechnical actions); (iii) GEO, to be used when considering the failure of the ground or where the strength of the soil provides significant resistance. Under normal situations, typically, ULS (STR) and SLS limit states should be considered.

2.6.2 Partial safety factors

The designer must consider whether sections of the complete reservoir may be empty when other sections are full and design each structural element for the maximum bending moments and forces that can occur due to (a) the hydrostatic pressures alone and (b) the lateral earth, groundwater and possible surcharge pressures or a combination of the pressures from (a) and (b).

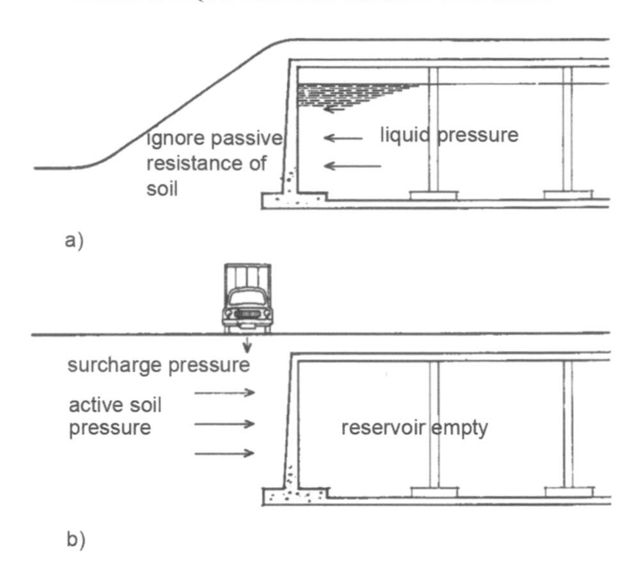

Figure 2.11 *Design loadings for external walls with soil fill (a) Reservoir full (b) Reservoir empty.*

As also directed by BS 8110 previously, the Eurocodes still design to the limit states by considering a combination of the permanent (dead) and variable (imposed) actions, where the characteristic actions are multiplied by an appropriate partial safety factor (psf). When designing a structural element for the ultimate limit state, it is necessary to use psfs (in conjunction with the characteristic actions) to provide the necessary margin against failure. The psfs take account of the likely variability of the loading and the consequences of failure.

For the case where the pressure is derived from the stored liquid alone (i.e. (a) above), the ULS (STR) operational safety factor, $\gamma_F = 1.2$, as provided in BS EN 1991-4 (Actions on Silos and Tanks). (Water = permanent action.). Under test, $\gamma_F = 1.0$.

For (b) above, it is usual to take the operational safety factor, $\gamma_F = 1.35$ for the permanent actions and $\gamma_Q = 1.5$ for the variable actions. Where there is more than one variable action, a multiplier, ψ_0, is applied to the variable action partial safety factor to reflect the statistical improbability that more than one variable action will be a maximum simultaneously with the others. There are two other multipliers. Multiplier ψ_1 is said to produce a frequent value of the load and multiplier ψ_2 a quasi-permanent value of the load. The frequent and quasi-permanent multipliers are typically used at the ULS where accidental actions are involved. The quasi-permanent multiplier can also be used to determine long-term effects such as creep and settlement. Numerical values of ψ_1 and ψ_2 are provided in BS EN 1990.

It should be noted that any of the combinations of permanent and variable actions discussed above relate to the magnitude of loads that could be present. The designer is still required to perform the structural analysis to determine the actual arrangement of these loads in the structure to create the most critical effect.

A few final design comments: as the roofs of partially buried and underground reservoirs are covered with a solar attenuating layer composed of soil or gravel, any imposed loads due to vehicles will be distributed before reaching the structural roof slab. In these circumstances, it will normally be appropriate to consider a single load/analysis case when designing the roof. Also, with respect to the roof of a reservoir, if the roof is monolithic with the walls, any thermal expansion of the roof may cause additional loading on the perimeter walls. BS EN 1991-1-5 (Thermal Actions) does provide some guidance on this effect; however, the guidance is more appropriate to bridge design. Research by the author is currently being performed (both in terms of monitoring a partially buried reinforced concrete service reservoir in North Yorkshire and full-scale laboratory testing) to quantify this type of thermal effect (Forth *et al.*, 2005; Muizzu, 2009; Forth, 2012). BS EN 1991-4 (Actions–Silos and Tanks) does state that stresses resulting from the restraint of thermal expansion can be ignored if the number of expansion cycles provides no risk of fatigue failure or cyclic plastic failure. Although the number of thermal cycles are relatively low, output from the research being performed by the author does suggest caution when designing monolithic roof to wall joints, even in buried or partially buried structures due to additional moments from thermal creep.

For a reservoir with height of wall, H and an operating depth of water, h, BS EN 1991-4 (Silos and Tanks) recommends that for operational conditions a partial safety factor, $\gamma_F = 1.2$ should be used (i.e. 1.2ρh, where ρ = density of the liquid) to calculate the design load at ULS. For accidental situations, it recommends that $\gamma_F = 1.0$; however, the full depth of the wall should be used (i.e. 1.0ρH).

2.7 Foundations

It is desirable that a liquid-retaining structure is founded on good uniform soil, so that differential settlements are avoided (Chapter 1). However, this desirable situation is not always obtainable. Variations in soil conditions must be considered and the degree of differential settlement estimated (Barnes, 2000). Joints may be used to allow a limited degree of articulation, but on sites with particularly non-uniform soil, it may be necessary to consider dividing the structure into completely separate sections. Alternatively, cut-and-fill techniques may be used to provide a uniform platform of material on which to found the structure.

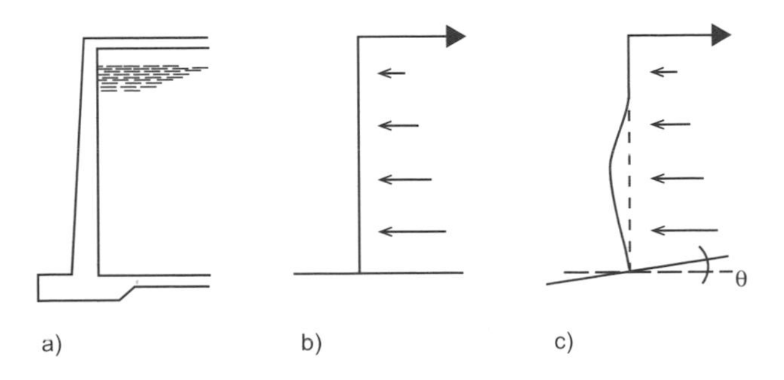

Figure 2.12 *Propped cantilever walls on a cohesive soil (a) Structure (b) Basic structural assumptions (c) Rotation due to soil movement.*

Soils that contain bands of peat or other very soft strata may not allow normal support without very large settlements, and piled foundations are required (Barnes, 2000; Manning, 1972).

The design of structures in areas of mining activity requires the provision of extra joints, the division of the whole structure into smaller units, or the use of rafts because of the potentially very large settlements. Prestressed tendons may be added to a normal reinforced concrete design to provide increased resistance to cracking when movement takes place (Davies, 1960; Melerski, 2000).

The use of partially buried cantilever walls depends on passive resistance to sliding between the base and the foundation soil. If the soil is inundated by groundwater, it may not be possible to develop the necessary resistance under the footing due to the high water pressures. In these circumstances, additional lateral restraint in terms of a toe would be required or alternatively in severe cases a cantilever design is not appropriate, and the overturning and sliding forces should be resisted by a system of beams balanced by the opposite wall, or by designing the wall to span horizontally if that is possible. When fully buried, it is reasonable to consider the resistance offered by the wall as well.

Walls that are designed as propped cantilevers, and where the roof structure can act as a tie, are often considered to have no rotation at the footing (Figure 2.12). However, the strain in a cohesive soil may allow some rotation and a redistribution of forces and moments.

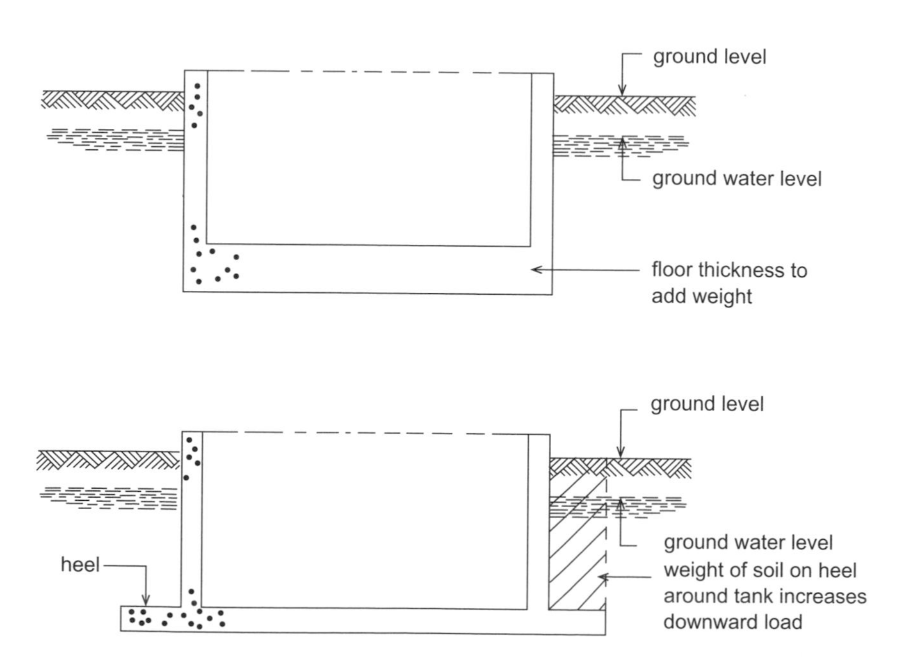

Figure 2.13 *Methods of preventing flotation (a) Additional dead weight (b) Provision of a heel.*

2.8 Flotation

An empty tank constructed in water-bearing soil will tend to move upwards in the ground, or float. The ability of the structure to resist this uplift can be checked by comparing the permanent stabilising actions (i.e. self weight and side friction) to the permanent and variable destabilising actions from the groundwater and possibly other sources. Simplistically, the designer should ensure that any tendency towards uplift must be counteracted by ensuring that the weight of the empty tank structure is greater than the uplift equal to the weight of the groundwater displaced by the tank.

The extent to which the stabilising actions must be greater than the destabilising actions is defined in BS EN 1997-1:2004 (Eurocode 7). This code guides the designer in the assessment of the stability of the structure against hydraulic uplift. BS EN 1997-1 (more specifically its National Annex) stipulates the value of the partial safety factors to be used when checking the uplift limit state (UPL) for buoyancy / flotation. The partial safety factor to be applied to the permanent stabilising action of the water is 0.9 ($\gamma_{G;stb} = 0.9$, where stb = stabilising) and the partial safety factor for the destabilising action is 1.1 ($\gamma_{G;dst} = 1.1$, where dst = destabilising).

Practically, the weight of the tank may be increased by thickening the floor or by providing a heel on the perimeter of the floor to mobilise extra weight from the external soil (Figure 2.13). Whichever method is adopted, the floor must be designed against the uplift due to the groundwater pressure. In calculating the weight of the soil over the heel, it is important to realise that the soil is submerged in the groundwater. The effective density of the soil is therefore reduced. If the floor is thickened, it is possible to construct it in two separate layers connected together by ties. This has the advantage that reduced thermal reinforcement appropriate to the upper thickness may be used.

The designer should consider conditions during construction, in addition to the final condition, and specify a construction sequence to ensure that the structure is stable at each phase of construction.

Chapter 6 provides a design example that illustrates the use of the basic design materials and parameters presented in this chapter, including a UPL design check for stability against hydraulic uplift.

Note

[1] Typically, owing to the more stringent cover requirements of BS EN 1992-1-1 and BS 8500, the required cover has increased compared to that required previously in BS 8110 and BS 8007. This has implications on the calculated crack widths and, where the crack spacing is controlled by the reinforcement, the steel area required to control these crack widths. Cracking is discussed more in Chapters 3 and 5.

Chapter 3
Design of reinforced concrete

3.1 General

The basic design philosophy of liquid-retaining structures is discussed in Chapter 2. In this chapter, detailed design methods are described to ensure compliance with the basic requirements of strength and serviceability.

In contrast with normal structural design, where strength is the basic consideration, for liquid-retaining structures it is found that serviceability considerations control the design. The procedure is therefore:

(i) estimate concrete member sizes;
(ii) calculate the reinforcement required to limit the design crack widths to the required value;
(iii) check strength;
(iv) check other limit states;
(v) repeat as necessary.

The calculation of crack widths in a member subjected to flexural loading can be carried out once the overall thickness and the quantity of reinforcement have been determined.

3.2 Wall thickness

3.2.1 Considerations

All liquid-retaining structures include wall elements to contain the liquid, and it is necessary to commence the design by estimating the overall wall thickness in relation to the height. The overall thickness of a wall should be no greater than necessary, as extra thickness will cause higher thermal stresses when the concrete is hardening.

The principal factors that govern the wall thickness are:

(i) ease of construction;
(ii) structural arrangement;
(iii) avoidance of excessive deflections;
(iv) adequate strength;
(v) avoidance of excessive crack widths.

The first estimate of minimum section thickness is conveniently made by considering (i), (ii) and (iii).

It will be found that a wall thickness of about 1/10 of the span is appropriate for a simple cantilever (Table 3.1), and somewhat less than this for a wall that is restrained on more than one edge. Each factor is discussed in the following sections.

3.2.2 Ease of construction

If a wall is too thin in relation to its height, it will be difficult for the concrete to be placed in position and properly compacted. As this is a prime requirement for liquid-retaining structures, it is essential to consider the method of construction when preparing the design. It is usual to cast walls up to about 8 metres high in one operation (note for panels over 7 m span, an adjustment must be made to the limiting span / depth ratios–see Section 3.2.5), and to enable this to be successfully carried out, the minimum thickness of a wall over 2 metres high should be not less than 250–300 mm. Walls less than 2 metres high may have a minimum thickness of 200 mm. A wall thickness less than 200 mm is not normally possible, as the necessary four layers of reinforcement cannot be accommodated with the appropriate concrete cover on each face of the wall. The wall may taper in thickness with height in order to save materials. Setting out is facilitated if the taper is uniform over the whole height of the wall (Figure 3.1).

3.2.3 Structural arrangement

Lateral pressure on a wall slab is resisted by a combination of bending moments and shear forces carrying the applied loads to the supports. The simplest situation is where the wall is a simple cantilever, with the maximum shear force and bending moment at

Table 3.1 *Approximate minimum thickness h (mm) of R. C. Cantilever wall subjected to water pressure.*

Height of wall (m)	*Minimum wall thickness h (mm)*
8	800
6	700
4	450
2	250

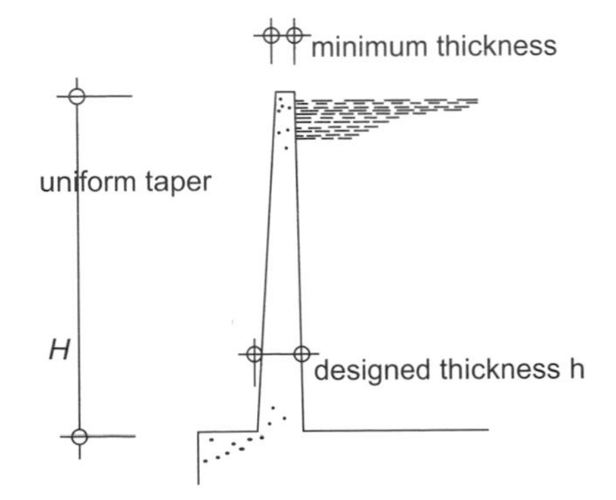

Figure 3.1 *Typical section through a wall.*

the base. This situation will require the thickest wall section as the bending moment is comparatively large. The most favourable arrangement is where a wall panel is held at all four edges and may be structurally continuous along the edges. In this case, the slab spans in two directions and in each direction there may be positive and negative moments. Each of the moments will be appreciably less than in the case of the simple cantilever, and hence a thinner wall is possible with less reinforcement to control cracking where the wall spans in two directions. The particular structural arrangement that is appropriate for a given design will depend on the relative spans in each direction and whether movement joints are required at any of the sides of the panel.

3.2.4 Shear resistance of reinforced concrete

Theoretically, shear in reinforced concrete flexural members is resisted by a combination of four factors:

(i) concrete in the compression zone;
(ii) dowell action of main reinforcement;
(iii) aggregate interlock across flexural (tension) cracks;
(iv) shear link reinforcement.

Eurocode 2 respects the above theory and as such, the shear stress depends on the concrete strength, effective depth and tension steel ratio. As before in BS 8110, the recommended design guidance in BS EN 1992-1-1 is (i) that there is a shear stress below which only minimum shear reinforcement need be provided (shear reinforcement is provided in all structural elements) and (ii) the design shear stress should be less than the shear capacity of the section.

However, there is a subtle dissimilarity between the approach in BS EN 1992-1-1 and that presented previously in BS 8110. In the former, there are effectively three stages in the design for shear. The first stage is to determine the capacity of the concrete alone. Should this capacity not be sufficient to resist the design shear force, the steel required to resist the designed shear is then determined without any consideration of the concrete's shear capacity (stage 2). Effectively, for the majority of structural beams, the shear capacity of the member will be calculated based only on the steel and ignoring the contribution of the shear capacity of the concrete. Stage 3 determines the specific area and spacing of the shear reinforcement.

It is inconvenient to use shear reinforcement in slabs because it is difficult to fix and it impedes the placing of the concrete. It is actually an inefficient use of steel. Therefore, in water-retaining structures, where the common element is a slab, shear design is performed by ensuring that the shear capacity of the concrete exceeds that of the (applied) design shear force, i.e. stage 1 above. (Stages 2 and 3 will not, therefore, be specifically discussed, although some reference will be made to the theory on which they are based. Details of stages 2 and 3 are adequately covered in the code and reference can also be made to several general texts on reinforced concrete (The Concrete Centre, 2005).)

According to BS EN 1992-1-1, the concrete shear force capacity, $V_{Rd,c}$ is given as:

$$V_{Rd,c} = b_w d\,[(0.18/\gamma_c)\,k\,(100\rho_1 f_{ck})^{1/3} + 0.15\sigma_{cp}] \text{ (units are N)} \qquad (3.1)$$

where

$(0.18/\gamma_c) = C_{Rd.c}$ where $\gamma_c = 1.5$ (partial factor for concrete)
$k = (1 + (200/d)^{1/2})$ 2.0 (with d expressed in mm)
$\rho_1 = A_{s1} / b_w d \leq 0.02$ where A_{s1} = the area of tension reinforcement that extends beyond the section being considered by at least a full anchorage length plus one effective depth, d
σ_{cp} is only included if there are axial forces within the member (discussed later)

In recognition of the fact that a member still possesses some shear strength even without any reinforcement, the minimum value for the concrete shear force capacity is:

$$V_{Rd,c} = [0.035k^{3/2} f_{ck}^{\ 1/2}]\, b_w d \text{ (units in N)} \qquad (3.2)$$

In order to ensure that there is sufficient capacity in the concrete, the wall thickness should be adjusted to suit such that the concrete shear force capacity, $V_{Rd,c}$ exceeds the applied shear force, V_{Ed}. Alternatively, the concrete shear stress capacity, $v_{Rd,c}$ must exceed the applied shear stress, v_{Ed}, where $v_{Ed} = V_{Ed} / 0.9\, b_w d$. (Note: 0.9 is only relevant when sections are being designed using the Variable Strut Inclination Method–see below.) The values for $C_{Rd,c}$, k_1 and v_{min} (where $v_{min} = [0.035k^{3/2} f_{ck}^{\ 1/2}]$) are provided in the National Annex. The National Annex also provides guidance for cases where the concrete strength classes are higher than C50/60. Whereas in BS 8110 the design shear stress was limited to the lesser of $0.8\sqrt{f_{cu}}$ or 5 N/mm², BS EN 1992-1-1 recommends that the applied shear force V_{Ed} should always satisfy the condition:

$$V_{Ed} \leq 0.5\, b_w d\, v f_{cd} \qquad (3.2a)$$

where $v = 0.6\,[1 - f_{ck}/250]$ (f_{ck} in MPa)

From Eq. (3.1) above, it is clear that the capacity of the concrete to resist shear is influenced by the longitudinal tension steel. It is, therefore, reasonable to expect that the applied shear force, V_{Ed} will cause an additional force in the tension steel and this needs to be considered in the design. However, BS EN 1992-1-1 does not require this check when designing members that **DO NOT** require design shear reinforcement. However, it does specify this check when designing members that **DO** require shear reinforcement. This additional longitudinal tension force, ΔF_{td} for sections reinforced with vertical links (i.e. links perpendicular to the horizontal or longitudinal axis of the section) is defined:

$$\Delta F_{td} = 0.5 V_{Ed} \cot\theta \qquad (3.3)$$

where θ is the angle between the concrete compression strut and the beam axis perpendicular to the shear force ($1 \leq \cot\theta \leq 2.5$; $22° \leq \theta \leq 45°$).

Owing to the method of design introduced in BS EN 1992-1-1 (the Variable Strut Inclination Method), it is easy to see how the compressive force in the inclined concrete strut needs to be balanced by a horizontal component tension force and therefore why the code requires the designer to consider this tension force due to the action of shear in the design of the tension steel. The code actually specifies that only half of this horizontal tension force is carried by the reinforcement in the tension zone. In all probability, the additional tension force due to shear will not be significant; it is unlikely that the required additional steel area when added to the area of bending steel will be greater than the area of steel already determined to satisfy the serviceability limit state. In fact, the additional force could probably be resisted by modifying the detailing of the steel (i.e. increasing the curtailment lengths of the tension reinforcement). However, the authors recommend that it would be good practice to perform the check (see examples in Chapter 6) and that θ (in Eq. (3.3) above) should be taken as 45° (i.e. cotθ = 2.5) to be conservative.

Whereas in BS 8110, values of design concrete shear stress were tabulated in terms of percentage area of tension steel and effective depth for a concrete of 25 MPa strength, BS EN 1992-1-1 does not provide such guidance. However, an equivalent table can be derived from Eqs (3.1) and (3.2) above and this is presented as Table 3.2 below. The table provides values of $v_{Rd,c}$ for slabs constructed with C30/35 concrete and without axial loads.

A comparison between the guidance provided in BS 8110 and the current BS EN 1992-1-1 shows that overall, the latter permits a lower shear stress before shear reinforcement is required (Moss and Webster, 2004). However, due to the minimum shear stress that can be carried according to BS EN 1992-1-1, the allowable shear stresses in this code tend to be higher for low reinforcement percentages (this difference is more obvious, the higher the strength of the concrete). It must be remembered that the theoretical behaviour of reinforced concrete in shear is difficult to analyse because of its complexity. The guidance presented in BS EN 1992-1-1, as was the case with BS 8110, is derived empirically from many experimental investigations. The differences between the current code and the old BS 8110, allowing for the new design method introduced in BS EN 1992-1-1, in many ways simply represent the additional test data that have been referenced, that were not considered or were not available when BS 8110 was drafted. A review by Collins *et al.* (2008) has still raised concerns over the ability of BS EN 1992-1-1 to safely predict the shear strength of members without links.

Shear with axial load

Equation (3.1) above for the design shear force resistance of members not requiring shear reinforcement (Expression 6.2a in BS EN 1992-1-1) also includes a term for

Table 3.2 *Shear force resistance of members without shear reinforcement,* $V_{Rd,c}$ *in* kN *(Class C30/35 concrete).*

$\rho_l = A_s/bd$	*Effective depth, d* (mm)										
	<200	*225*	*250*	*275*	*300*	*350*	*400*	*450*	*500*	*600*	*750*
0.25	108	117	125	132	141	157.5	172	184.5	200	228	270
0.50	118	128.25	140	151.25	162	182	204	220.5	240	282	337.5
0.75	136	148.5	160	173.25	186	206.5	232	252	275	318	382.5
1.00	150	162	177.5	189.75	204	227.5	256	279	305	354	427.5
1.25	160	175.5	190	203.5	219	248.5	276	301.5	330	378	457.5
1.50	170	186.75	202.5	217.25	234	262.5	292	319.5	350	402	487.5
1.75	180	197.75	212.5	228.25	246	276.5	308	337.5	365	426	510
2.00	188	204.75	222.5	239.25	255	287	320	351	385	444	532.5
2.50	188	204.75	222.5	239.25	255	287	320	351	385	444	432.5
k	2.000	1.943	1.894	1.853	1.816	1.756	1.707	1.667	1.632	1.577	1.516

cases where the member is subjected to an axially applied load as well as a bending moment. The axial stress, σ_{cp} is defined as:

$$\sigma_{cp} = N_{Ed} / A_c < 0.2 f_{cd} \tag{3.4}$$

where

N_{Ed} is the axial force in the cross section due to loading (+ for compression, − for tension) (in N). The influence of imposed deformations may be ignored
A_c is the area of concrete cross section (mm^2)
f_{cd} is the design value of concrete compressive strength

The inclusion of this axial stress parameter is in harmony with other International codes and Cl 3.4.5.12 of BS 8110. However, whilst it seems applicable to cases of prestressing, where the axial stress is compressive, or for other 'normal' cases producing axial compression, its application and modification of allowable design shear resistance in cases where axial tensions are present has always been contentious.

From Eq. (3.1) it can be seen that if the axial force is compressive, then $V_{Rd,c}$ is enhanced; there is no change to the tension steel and the axial force is resisted by the concrete. The enhancement of the allowable shear resistance is reasonable as the compression enhances two of the factors resisting shear, i.e. it would increase the compression force in the compression zone, and it could possibly enlarge the compression zone, improving resistance to cracking (and perhaps reduce existing crack lengths/size) and improving aggregate interlock.

This may be critical for water-retaining structures, as typically these type of structures are designed as rectangular boxes with high tensions being developed in the one or two-way spanning wall elements. Consequently, the combined case of shear with axial tension is prevalent at the corners. Equation (3.1) suggests that an axial tension force will effectively reduce the allowable shear resistance of the section possibly leading to shear reinforcement being required or an increase in section depth. An alternative may be to increase the tension steel to allow for this additional stress. With the axial tension stress being taken by the flexural steel, there would be no reason to reduce $V_{Rd,c}$. Again, in all probability, the additional steel area needed to resist the axial tension added to the flexural steel area would still not exceed the steel area required to resist the serviceability conditions. The fact is, by adopting this approach it may be the case that in most cases the designer does not need to consider axial tension. Again, it would be good practice as a designer to check the impact of the axial tension; however, this may be a way forward without detrimentally affecting $V_{Rd,c}$.

As was the case for BS 8110, the guidance provided in BS EN 1992-1-1 with respect to the case of combined shear and axial tension is empirically derived. Unfortunately, the derivation is based on the results of a very small number of investigations. Where this has been investigated, the specimens tested are more likely to be beam elements and most already contain shear links, although some tests were performed on beams without link reinforcement. Observations from these tests suggested that axial tension loads had little effect on the shear strength of the beams and that axial tension only became influential when samples possessed small shear span / effective depth ratios (a/d = 1.96), which is largely irrelevant to water-retaining structures. These findings appear to support the proposed alternative above that infers that axial tension is not critical. However, there is clearly the need for further research with respect to combined shear and tension, particularly in slabs and monolithic indeterminate slab joints.

The maximum shear force in a cantilever occurs at the foot of the wall immediately above the base, and the shear stress in the concrete is also a maximum at this level. However, in Cl 6.2.1 (8) of BS EN 1992-1-1 it is recommended that (i) conservatively, the design shear force need not be checked at a distance of less than the effective depth, d of the wall from the base level or (ii) in cases where the load is applied close to the support (see Figure 3.2), the applied shear force, V_{Ed} (calculated within a distance $0.5d$ $a_v \leq 2d$) may be reduced by a factor $\beta = a_v / 2d$ and the design shear resistance of the member, $V_{Rd,c}$ checked against this reduced V_{Ed}. Alternatively, as presented by Narayanan and Beeby (2005), the design shear resistance can be enhanced by the inverse of β for members where the load is applied up to $2d$ from the face of the support as:

$$V_{Rd,c} = b_w d\,[(0.18/\gamma_c)\,k\,(100\rho_1 f_{ck})^{1/3}\,(2d/a_v) + 0.15\sigma_{cp}] \qquad (3.5)$$

where a_v is the distance from the face of the support to the face of the load.

This is provided that the longitudinal reinforcement is fully anchored at the support and is in recognition of the fact that a significant proportion of the applied load would be carried through to the base support due to the angle of shear at this location. The section of the wall between the base and the face of the load (a maximum distance of $2d$ from the base) need not be checked.

Consider the free cantilever wall of the uniform tapered section subjected to water pressure shown in Figure 3.2.

H = height of water (m);
γ_w = density of water (kN/m³);
h = maximum overall thickness of section (mm);
d = maximum effective depth of section (mm);
a = depth of centre of tension steel from face of concrete (mm); the overall thickness $h = d + a$;
γ_f = partial safety factor for action.

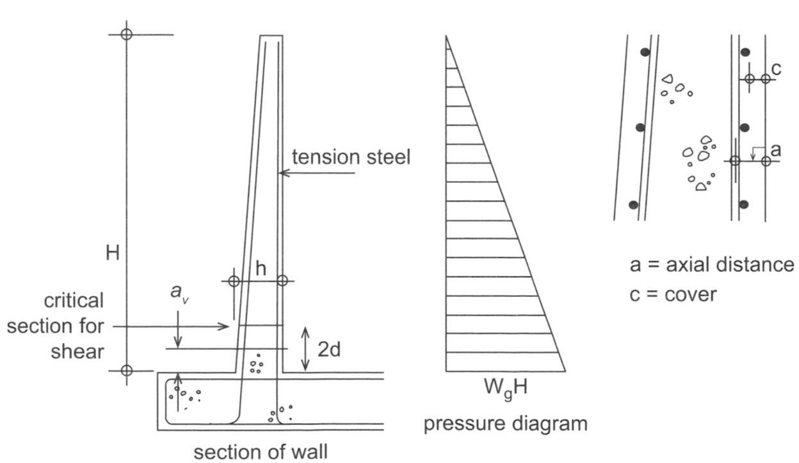

Figure 3.2 *Cantilever wall subjected to water pressure.*

The applied ultimate shear force at the critical section of the cantilever is:

$$V_{Ed} = 0.5\, \gamma_w\, \gamma_f\, (H - 2d)^2$$

(Note: d in the above equation must be in m and not mm for consistency of units.) Therefore, the applied ultimate shear stress on the section, $v_{Ed} = V_{Ed} / b_w d$

The distance from the face of the concrete to the centre of the tension steel, a, varies according to bar size and cover. Allowance should be made for any taper on the section. Assuming that the concrete cover is 40 mm and the bar size is 16 mm, the value of a is equal to 40 + (1.5 × 16) or about 65 mm. (Distribution reinforcement should be in the outer layer where it is more effective.) The required section thickness h may be calculated from given values of applied shear force and design shear stress resistance to ensure that no shear reinforcement is required. An example of the calculation follows.

Example 3.1 *Calculation of wall thickness*

Consider a cantilever wall of height H (i.e. equal to the water height) subject to water pressure where the height $H = 6.0$ m.
Density of water $= \gamma_w = 10$ kN/m³. The Partial safety factor $\gamma_f = 1.2$.
Assume tension reinforcement ratio, $\rho_1 = 100A_s / b_w d = 0.5\%$
Maximum applied ultimate shear force at base level:
$V_{Ed} = 0.5 \times 10 \times 1.2 \times 6.0^2 = 216$ kN per m run

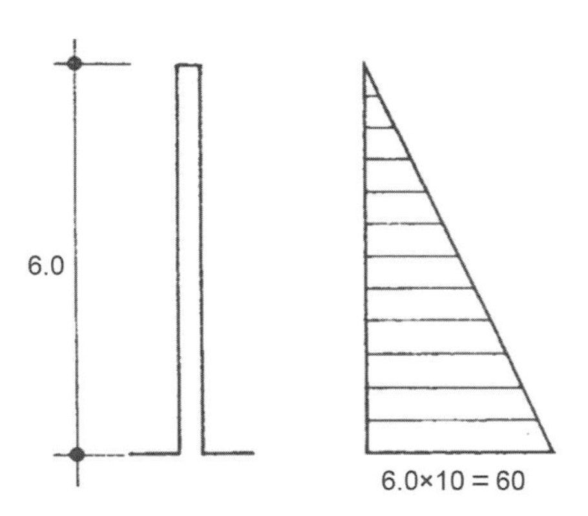

Assuming a wall thickness of h = 700 mm (see Table 3.1) and an effective depth of 600 mm, the critical section for shear will be at a level of 1200 mm above the base, and the critical applied shear force will be

$$V_{Ed} = 216 \times (4.8 / 6.0)^2 = 138 \text{ kN/ per m run}$$

Referring to Table 3.2, for a C30/35 concrete and 0.5% reinforcement ratio, the minimum effective depth required to resist an ultimate applied shear force of 138 kN per m width may be estimated as 250 mm and, therefore, the overall wall thickness

$$h = d + a = 250 + 65 = 315 \text{ mm}$$

From shear considerations alone, the wall should have a minimum thickness of approximately 325 mm.

It would then be necessary to recheck the calculation using the new value of *d;* however, the required thickness for a 6 m high wall of approximately 600 to 700 mm easily satisfies shear. In this case, shear is not critical.

3.2.5 Deflection

The lateral deflection of a cantilever wall that is proportioned according to the rules suggested in this chapter is likely to be no more than about 30 mm. A wall that is restrained by connection to a roof slab or by lateral walls will clearly deflect even less. Deflection of this magnitude will have no effect on the containment of liquid and, unless there is a roof slab supported by the wall with a sliding joint, there is no need to consider the amount of deflection. If pipes or other apparatus pass through a wall that may itself move slightly under load, the pipes must be arranged to be sufficiently flexible to allow for this movement.

Eurocode 2 allows members to have stiffness defined in terms of span/effective depth ratios as an alternative to calculating deflections. These values apply equally to normal and liquid-retaining structures. Typical values are given in Table 7.4N in BS EN 1992-1-1 for both lightly ($\rho = 0.5\%$) and highly stressed ($\rho = 1.5\%$) concrete and apply to a range of concrete compressive strengths. Interpolation between these values of ρ is allowed. The values are based on limiting the deflection of a slab, which is subjected to quasi-permanent loads, to span/250 and assumes that the member is constant in depth and that the loading is uniform. The span/250 ratio provides a limit to the total deflection and is much more relevant for this type of structure. BS EN 1992-1-1 also suggests that the designer consider a long-term deflection limit of span/500, normally appropriate for consideration of quasi-permanent loads after construction (i.e. the deflection occurring after the installation of finishes and partitions); however, this will not, in the majority of cases, be relevant here. Deflection limits are discussed further below. In the case of a vertical cantilever wall subjected to liquid pressure, the loading will be of triangular distribution and the wall section may be tapered. If the values in BS EN 1992-1-1 are used as the basis for calculating the effective depth of the member, a slightly conservative design will result. Allowance may be made for the effect of the triangular load distribution by increasing the basic allowable ratio for a cantilever by 25%. (This is based on a comparison of deflection coefficients.)

Table 3.3 is based on the recommendations of BS EN 1992-1-1 UK National Annex and provides span/effective depth values for lightly stressed concrete (when the steel ratio = 0.5%), which is more typical for the case of slabs. The limiting span/effective depth ratios can also be estimated using Expressions 7.16(a) and 7.16(b) of BS EN 1992-1-1. (Expressions 7.16(a) and (b) provide in effect a permissible slenderness ratio; which one is used is dependent on whether the design tension steel is $\leq$ or $>$ a reference steel ratio. The output from (a) or (b) is compared to the actual slenderness ratio (actual span/effective depth) to assess conformity.) From these expressions it can clearly be seen how the limiting ratios are dependent on the compressive strength of the concrete, the area of tensile and compression steel and the different structural system (or element), which is defined as K in the code and shown in Table 3.3. These expressions have been derived assuming that the steel stress at a cracked section and at SLS is 310 MPa (f_{yk} = 500 MPa). Where other stress levels are used, the output from Expression 7.16 of the code (or the basic ratios provided in Table 3.3) should be multiplied by 310 /σ_s, where σ_s = the tensile stress at midspan (or support for a cantilever) under the design load at SLS. It is conservative to obtain σ_s using the ratio of area of steel required for ultimate limit state, $A_{s,\,req}$ to area of steel provided, $A_{s,\,prov}$ as:

$$310 / \sigma_s = (500 / f_{yk}) (A_{s,\,req} / A_{s,\,prov}) \tag{3.6}$$

The maximum adjustment when considering other stress levels is 1.5 (i.e. $A_{s,prov} \leq 1.5\ A_{s,req}$). [BS EN 1992-1-1 NA]

An additional factor, which can also be applied to the output of Expression 7.16(a) and (b) (or the basic ratios) is applicable when the span of the slab (not flat slab) exceeds 7 m. When the span is greater than 7 m, Expression 7.16 should be multiplied by 7 / l_{eff} where the effective length, l_{eff} is defined in clause 5.3.2.2 (1) of BS EN 1992-1-1. Recently, modifications to the span to depth rules, i.e. to account for load ratio/history; slab thickness; and reinforcement ratio have been presented by Vollum (2009). His investigation has shown that potentially the span to depth rules presented in BS EN 1992-1-1 are more versatile (than the original rules developed by Beeby and incorporated in BS 8110) but that they can provide unreliable results as they are currently formulated. The Concrete Centre is currently developing an improved set of rules which potentially could also lead to more economic solutions.

As mentioned above, there is the option to calculate the actual deflections, and BS EN 1992-1-1 provides guidance on how this should be achieved. However, as will be discussed below, the authors have a number of concerns with respect to the deflection control limits and the assumed values for shrinkage and creep curvature. It is also important to note that the theory presented in BS EN 1992-1-1, specifically relating to cracked sections, is not based on any experimental research.

Calculation of deflection

In the design for deflection control, two limits are given in BS EN 1992-1-1. There is a limit to the total deflection of span/250 and a limit to the deflection occurring after the installation of finishes and partitions of span/500. In checking the limit after the

Table 3.3 *Basic ratios of span/effective depth for reinforced concrete members without axial compression.*

System	*K*	*Highly stressed* $\rho = 1.5\%$	*Lightly stressed* $\rho = 0.5\%$
simply supported beam, one- or two-way spanning simply supported slab	1.0	14	20
end span of continuous beam or one-way continuous slab or two-way spanning slab continuous over one long side	1.3	18	26
interior span of beam or one-way or two- way spanning slab	1.5	20	30
slab supported on columns without beams (flat slab) (based on longer span)	1.2	17	24
cantilever	0.4	6	8

installations of finishes and partitions it is normal to assume that the short-term deflection under the dead weight of the structure has occurred and that the deflection occurring after the installation of finishes and partitions consists of:

(i) a short-term increase in deflection resulting from an increase in loading due to application of the imposed or variable loads;
(ii) an increase in deformation due to creep and loss of tension stiffening over time;
(iii) an increase in deflection due to shrinkage.

Figure 3.3 shows, for what seems reasonable practical conditions, the relative magnitudes of these components. In the calculations, the procedures and values for estimating creep and shrinkage have been taken from BS EN 1992-1-1. The vertical axis gives values of the effective depth multiplied by the curvature, which is a convenient non-dimensional parameter defining bending deformation. This is directly proportional to the deflection. Inspection of this figure shows that, for the particular case considered, the increment in deflection after installation of the finishes and partitions is more than half the total deflection and hence the span/500 limit will be the governing factor in checking the deflection rather than the actual total deflection. Of this increment in deflection, roughly a third is calculated to be due to shrinkage. The importance of shrinkage may come as a surprise, though it has been noted by Alexander (2002) who discusses the great differences between the predictions of various current design formulae.

Shrinkage causes deflection in any member that is unsymmetrically reinforced because the reinforcement, which does not shrink, restrains the surrounding concrete from shrinking. In, for example, a singly reinforced member, the reinforcement

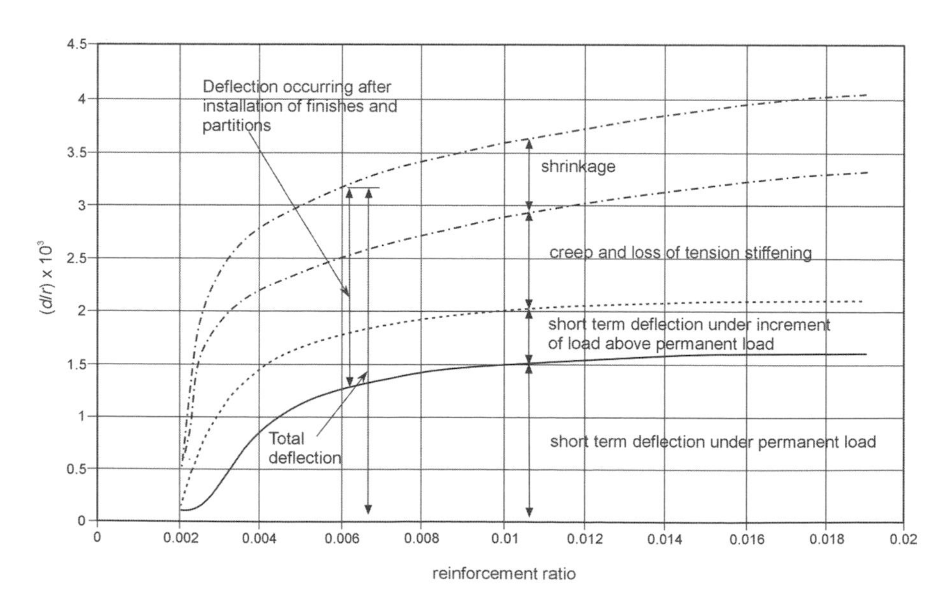

Figure 3.3 *Components of long-term deformation.*

restrains the shrinkage near the reinforced face but not at the opposite face. The result is that the unreinforced face shortens relative to the reinforced face, which results in a curvature and hence a deflection. A formula for the calculation of the deformation due to shrinkage in an uncracked beam was developed by Hobbs (1979), though the basic analytical approach had been considered in the USA a considerable number of years earlier (Parrott, 1979). This formula appears in BS EN 1992-1-1 and is:

$$1/r_{cs} = \varepsilon_{cs}\alpha_e S/I$$

where

- $1/r_{cs}$ = the curvature at the section considered due to restrained shrinkage
- ε_{cs} = the free shrinkage of an unrestrained and unreinforced concrete section of the same cross section
- α_e = the modular ratio ($=E_s/E_{c,\,eff}$)
- S = the first moment of area of the reinforcement about the section centroid
- I = the second moment of area of the section about its centroid

The section centroid is calculated allowing for the effect of the reinforcement within the uncracked section.

This formula can easily be derived on the basis of equilibrium of forces and compatibility of strains. There is little argument about its correctness for uncracked sections; indeed, Hobbs tested the equation against results from a considerable number of uncracked beams and concluded that the method was the most reliable of those available at the time. In any case, the calculated curvatures are generally small. In Figure 3.3, sections that are likely to be uncracked throughout their service life are those with reinforcement ratios below about 0.0025. What remains far more uncertain is the effect of shrinkage on a cracked section. BS EN 1992-1-1 simply proposes that S and I are calculated for the cracked section rather than the uncracked section. This solution was arrived at without any experimental confirmation of its validity and, as may be seen from Figure 3.3, calculating the curvature in this way results in a roughly four-fold increase in the calculated deformation once the section is considered to be cracked.

BS EN 1992-1-1 does acknowledge that the deformation of a cracked member is actually a combination of the behaviour of an uncracked and cracked section and that the actual deformation will be intermediate between these two limits. For members subjected mainly to flexure, the predicted behaviour is given by Expression 7.18:

$$a = \zeta a_{II} + (1 - \zeta)a_I \qquad (3.7)$$

where a is the deformation parameter considered, which may be, for example, a strain, a curvature or a rotation; a_I, a_{II} are the values of the parameter calculated for the uncracked and fully cracked conditions, respectively; ζ is a distribution coefficient (allowing for tension stiffening at a section) given by:

$$\zeta = 1 - \beta[\sigma_{sr} / \sigma_s]^2 \qquad (3.8)$$

where

- ζ = 0 for uncracked sections;
- β is a coefficient taking account of the influence of the duration of loading or of repeated loading on the average strain;

β = 1.0 for a single short-term loading and β = 0.5 for sustained loads or many cycles of repeated loading. (For water-retaining structures, where the number and frequency of cycles is low (water levels in service reservoirs forming part of an efficiently supervised potable water system may only change drastically during maintenance and major peak demand periods) no further loss in tension stiffening may be expected. However, it should be noted that tension stiffening benefits can possibly be reduced, in some cases to zero, depending on the frequency of loading (Higgins *et al.*, 2013));

σ_s is the stress in the tension reinforcement calculated on the basis of a cracked section;

σ_{sr} is the stress in the tension reinforcement calculated on the basis of a cracked section under the loading conditions causing first cracking.

The difficulty in making a realistic assessment of the effects of shrinkage on a cracked beam is that it is hard to see how to separate the factors influencing long-term behaviour. Uncracked beams can simply be tested in an unloaded state but cracked beams can only be cracked by virtue of the presence of load. If a beam is tested that is loaded to above the cracking load then it is hard to see how to differentiate between changes in deformation resulting from creep, from loss of tension stiffening and from shrinkage. Also, since many of the parameters affecting creep are the same as those affecting shrinkage, attempts to compare mixes with different shrinkages will almost certainly also result in mixes with different creep characteristics. The lack of reliable test data in this area almost certainly arises from this difficulty in establishing unequivocally what deformations are specifically due to shrinkage and what deformations arise from other causes.

There is practical significance to the establishment of a realistic value for shrinkage deformation. If the shrinkage curvature was significantly smaller than currently calculated for a cracked member then, since the long-term increment in deformation would be substantially reduced, the span/500 limit could become generally irrelevant in design as the total deflection would always be the critical factor. Proposed changes to the treatment of long-term tension stiffening (Beeby and Scott, 2003) would lead to an increase in the immediate deflection under the permanent load and a decrease in the long-term increment, and add further strength to the possibility that the long-term increment in deflection may be proved to be generally irrelevant to design. As mentioned above, and accepted historically in the design of water-retaining structures, it is highly unlikely that the long-term ratio would be relevant for these types of structure and so this potential outcome would not be significant here. However, the fact that realistic values for shrinkage (i.e. what is the shrinkage curvature of a section which may or may not be symmetrically reinforced (or have steel at an equal distance from the centre line of the section) and which may or may not have uniform drying potential from each face) and to some degree creep deformation in cracked sections are not available could suggest that the span/250 ratio is somewhat conservative. Ongoing research by the author is in the process of successfully quantifying the shrinkage curvature of cracked reinforced concrete sections for the first time (Forth *et al.*, 2012; Scott *et al.*, 2011).

The discussion above deals with the deflection of cantilevers assuming a fixed base, but further lateral deflection may be caused by rotation of the base due to consolidation of the soil. This factor is of importance for high walls and relatively compressible ground. An estimation of the lateral deflection at the top of a wall due to base rotation may be made by considering the vertical displacements of the extremities of the foundation with the reservoir full, assuming that the wall and base are rigid, and subjected to rotation calculated from the differential soil consolidation at front and rear of the footing (Figure 3.4).

$$\text{Rotation } \phi = (a_1 - a_2) / B \quad \text{and } a_r = \phi H \tag{3.9}$$

The value must be added to the deflection due to the flexure of the wall calculated by

$$a_w = \gamma_f H^5 / 30EI$$

In this formula, H may be taken to the top of the base slab. Finally, the total deflection at the top of the wall

$$a = a_w + a_r \tag{3.10}$$

is compared with $H/250$ or any other requirement.

With a propped cantilever wall, deflection will not be critical, but the rotation of the base will alter the relation between the negative and positive moments in the wall. The moments may be calculated most easily using a computer program.

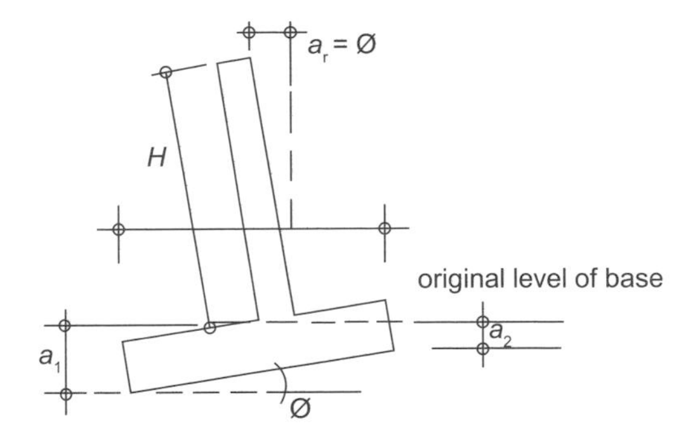

Figure 3.4 *Rotation of cantilever wall due to soil consolidation.*

3.3 Cracking

If a reinforced concrete slab is laterally loaded, the concrete on the side of the tension reinforcement will extend and, dependent on the magnitude of loading (other factors being equal), it will eventually crack as the load is increased. At the instant that a crack forms, it will have a positive width. Further increases in load will produce more cracks and a widening of existing cracks that have formed, ultimately leading to a stabilised crack pattern (although this is unlikely in practice), all the time increasing the stress in the reinforcement (Figure 3.5). For the same concrete section and load but with a greater quantity of reinforcement, the service stresses in the steel will be reduced, and the crack widths will potentially be narrower.

The applied load is fixed by the structural arrangement and, using limit state design, the designer has to choose values of slab thickness and reinforcement quantity to ensure that the crack widths under service loads are within the appropriate values given by the class of exposure (Chapter 2), and that the ultimate limit state is satisfied. There is no single design that will simultaneously exactly meet all the required criteria, and a number of different solutions are possible, even for a given value of design crack width.

The detailed methods of calculation for limit state design are considered in Sections 3.4 and 3.5. Where direct tensile forces are present in addition to flexural forces, the designer should consider which force system is predominant. In a vertical wall, some horizontal tension will be present, adjacent to lateral walls. In a circular deep tank, there will be almost entirely tensile forces and no flexure towards the top of the wall. When flexural forces are predominant, the allowance for the tensile forces may be made by adding to the calculated reinforcement resisting flexure, an extra quantity calculated by reference to the service stress in the flexural steel. If both flexure and direct tension are present to a significant degree, a calculation should be made using a modified strain diagram across the section to allow for the tensile force. The flexural crack width calculation may then be made. If a significant tensile force is present in a section, it is necessary to have reinforcement disposed in each face of the section in nearly equal quantities. Having regard to the avoidance of errors on site, it is good and sensible practice either to have equal steel arrangements in each face of a wall, or to have visibly distinct arrangements.

The precise calculation of the stress and strain diagrams for combined bending and tension results in a cubic equation of some complexity. Various designers' handbooks provide solutions using charts, and the alternative is to use a computer programme. In some circumstances, it is possible in the first instance to choose a section and

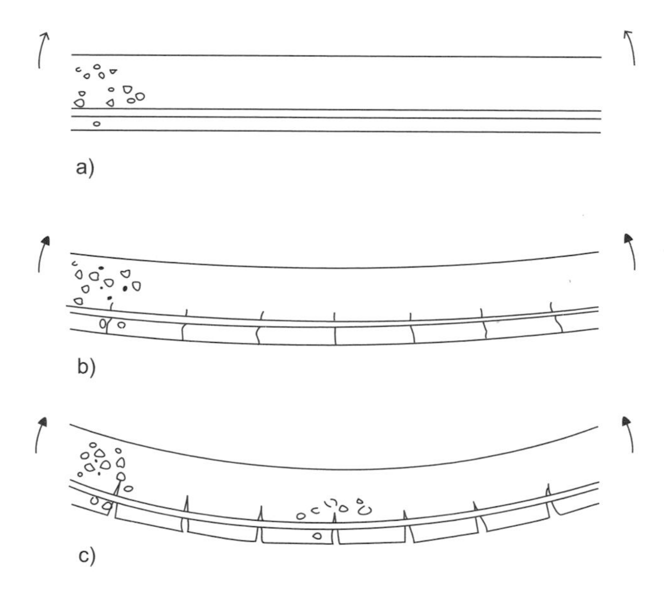

Figure 3.5 *Flexural cracking. (a) Concrete uncracked with low steel stress (b) Fine cracks and increased steel stress (c) Wide cracks and high steel stress.*

reinforcement by considering bending only, and then to modify the design using the formulae given in Section 3.6 in order to recalculate the strain from the depth of the neutral axis which is appropriate for the actual applied bending and tension. The results are then iterated until a satisfactory solution is obtained.

3.4 Calculation of crack widths due to flexure

3.4.1 Stress limitations in the concrete and steel

The limit state of cracking is satisfied by ensuring that the maximum calculated surface width of cracks is not greater than the specified value, depending on the degree of exposure of the member (see Chapter 2). To check the surface crack width, the following procedure is necessary:

- (i) calculate the service bending moment;
- (ii) calculate the depth of the neutral axis, lever arm and steel stress by elastic theory;
- (iii) calculate the average surface strain allowing for the stiffening effect of the concrete;
- (iv) calculate the crack spacing;
- (v) calculate the crack width.

The maximum service bending moment is calculated using characteristic loads with $\gamma_f = 1.0$. The calculation for a slab is based on a unit width of 1 metre.

The depth of the neutral axis x is calculated (see Section 3.8.1) using the usual assumptions for modular ratio design (Figure 3.6):

$$x / d = \alpha_e \rho \ \{(1 + (2 / \alpha_e \rho)^{1/2} - 1\} \tag{3.11}$$

where

α_e is the modular ratio = $E_s / E_{c,eff}$

$\rho = A_s / bd$

Alternative versions of this formula to calculate the depth of the neutral axis are available (see Chapter 6 Design calculations). Also, a similar but more complex formula may be used when compression reinforcement is present.

$E_{c,eff}$ (long-term or effective elastic modulus) above has traditionally in many cases been taken as half of the instantaneous or short-term modulus of elasticity of concrete, E_{cm}. The short-term modulus is given in Table 3.1 of BS EN 1992-1-1 for concrete strengths from 20 MPa to approximately 100 MPa (cube strength). However, it is now more correct and accurate to define the effective modulus by first using the nomograms provided in Figure 3.1 of BS EN 1992-1-1 to identify the creep coefficient, ϕ (∞, t_0) for a particular strength concrete; the effective modulus is then calculated from:

$$E_{c,eff} = E_{cm} / (1 + \phi\ (\infty, t_0)) \tag{3.12}$$

Figure 3.1 of BS EN 1992-1-1 relates the tangent modulus, E_c to the creep coefficient; E_c may be taken as 1.05 E_{cm} (see Cl 3.1.4 (2)). The values of E_{cm} given in Table 3.1 of BS EN 1992-1-1 are for applied compressive stresses of between 0 and $0.4f_{cm}$ and concretes containing quartzite aggregates. A 10% reduction in the values should be

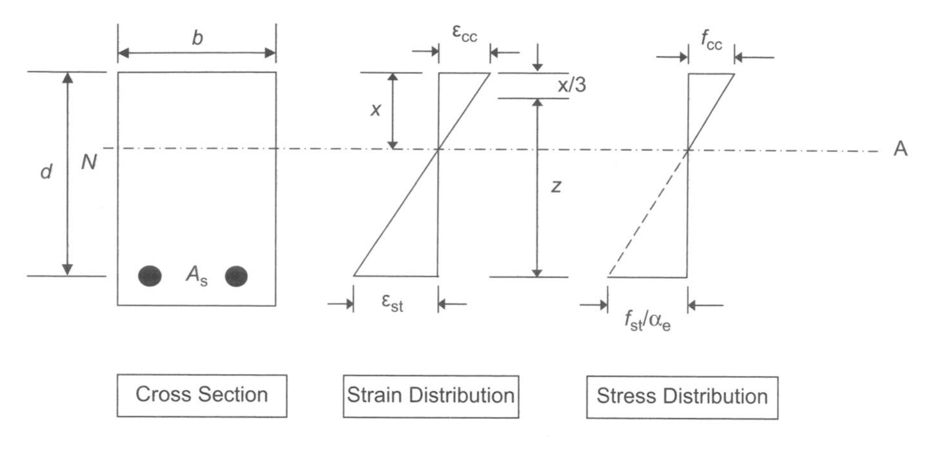

Figure 3.6 *Assumed stress and strain diagrams – cracked section – elastic design.*

applied if the concrete is manufactured using limestone aggregates (see Cl 3.1.3 (2)). (Note: The use of a long-term or effective modulus defined in this way is currently under investigation by the author. Long-term moments due to creep recorded in a monolithic roof-wall joint have been seen to be much greater than those predicted using a reduced or effective modulus (Forth, 2012; Muizzu, 2009).)

Once x has been determined from Eq (3.11) above, the lever arm, z can be found from

$$z = d - x / 3 \tag{3.13}$$

The tensile steel and concrete compressive stresses are then given using:

$$f_{st} = M_{sls} / zA_s$$

$$f_{cc} = 2M_{sls} / zbx$$

According to BS EN 1992-1-1 Cl 7.2 (3) and (5), in order to assume linear creep and for the crack width formula to be valid, the compressive stress in the concrete under quasi-permanent loads and the tensile stress in the steel under service conditions must be less than the following limiting values:

concrete: $k_2 f_{ck}$
steel: $k_3 f_{yk}$

where $k_2 = 0.45$ and $k_3 = 0.8$.

3.4.2 Flexural cracking

As with BS 8007, BS EN 1992-3 also provides guidance for the design of reinforcement to control early-age thermal and shrinkage cracking and flexural cracking. Early-age thermal cracking is considered in Chapter 5 (which also considers long-term cracking due to shrinkage). This section deals with flexural cracking.

To best illustrate the format of the guidance offered in BS EN 1992-3 it is worth recapping the approach presented in BS 8007. Previously, BS 8007 defined the design surface crack width for sections in flexure or combined flexure and tension where the depth of the neutral axis, x was between 0 and d (effective depth) as:

$$w = 3a_{cr}\varepsilon_m / [1 + 2((a_{cr} - c_{min}) / (h - x))] \quad (3.14)$$

where the average strain at the level that the cracking is being considered, $\varepsilon_m = \varepsilon_1 - \varepsilon_2$

ε_1 is the strain at the level considered (ignoring the stiffening effect of the concrete in the tension zone) (see Figure 3.7).

$$\text{Hence, } \varepsilon_1 = [(h-x) / (d-x)].f_{st} / E_s \quad (3.15)$$

and ε_2 is the strain due to the stiffening effect of concrete between cracks and this was defined in terms of limiting design surface crack widths of either 0.2 mm or 0.1 mm (BS 8007: Appendix B; Equations 2 and 3, respectively).

The hyperbola (Equation 3.14) describes the development of crack width both in terms of crack spacing and with distance away from the bar and is asymptotic to the maximum crack width, w_{lim}. When $a_{cr} = c_{min}$ (i.e. immediately over the bar), the crack width $w = 3c_{min}\varepsilon_m$.

In BS 8007, the format of the equation to calculate the crack width of members subjected to flexure is different to that provided to calculate crack width resulting from early-thermal and shrinkage movements. However, Equation 7.8 of BS EN 1992-1-1 (here as Equation 3.16) describes the crack width for both cases:

$$w_k = s_{r,max} (\varepsilon_{sm} - \varepsilon_{cm}) \quad (3.16)$$

where $s_{r,max}$ = is the maximum crack spacing;

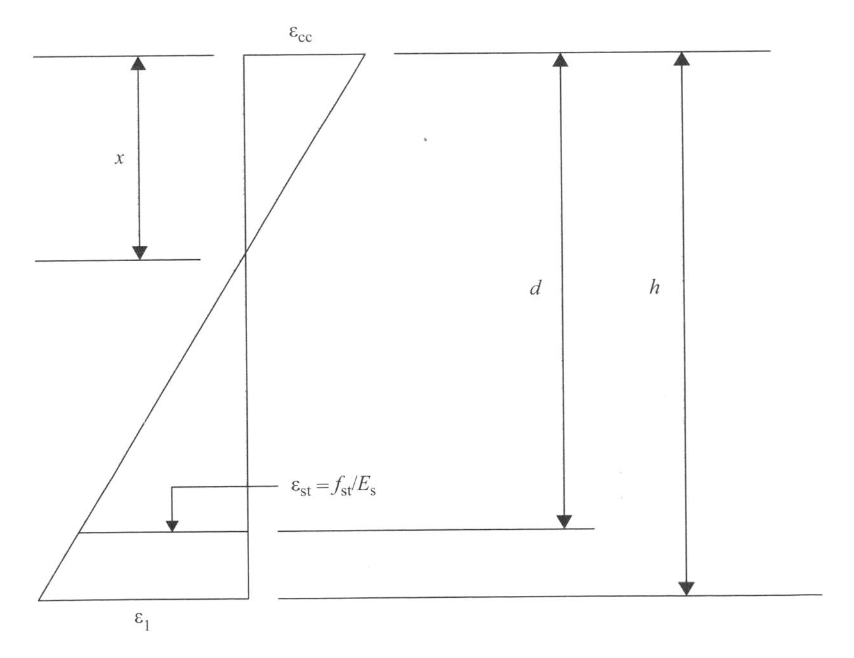

Figure 3.7 *Crack calculation – strain diagram.*

ε_{sm} = is the mean strain in the reinforcement under the relevant combination of loads, including the effects of imposed deformations and taking into account the effects of tension stiffening;

ε_{cm} = the mean strain in the concrete between cracks (often ignored).

It must be noted that the approach adopted in BS EN 1992-1-1 (as was the case with BS 8110) does not predict the maximum crack width but a crack width that in practice has a 20% chance of being exceeded (Beeby, 1979).

For the flexural case, $(\varepsilon_{sm} - \varepsilon_{cm})$ can be calculated using Equation 7.9 of BS EN 1992-1-1:

$$(\varepsilon_{sm} - \varepsilon_{cm}) = \sigma_s - k_t\,(f_{ct,eff} / \rho_{p,eff})(1 + \alpha_e \rho_{p,eff}) / E_s \geq 0.6\,(\sigma_s / E_s) \qquad (3.17)$$

where

σ_s = the stress in the tension reinforcement assuming a cracked section;
α_e = the modular ration E_s / E_{cm};
$\rho_{p,eff} = (A_s + \zeta_1{}^2 A_p') / A_{c,eff}$;
A_p' = 0 (reinforcement only; no pre or post-tensioned tendons);
$A_{c,eff}$ = the effective area of concrete surrounding the reinforcement of depth $h_{c,eff}$ (where $h_{c,eff}$ is defined in Figure 3.8);
ζ_1 = 0 (only consider if prestressing cables present);
k_t = a factor dependent on the duration of load (0.6 for short term; 0.4 for long term).

The maximum final crack spacing is calculated from Expression 7.11 of BS EN 1992-1-1:

$$s_{r,max} = k_3 c + k_1 k_2 k_4 \phi / \rho_{p,eff} \qquad (3.18)$$

where

k_3 and k_4 are found in the National Annex and are 3.4 and 0.425, respectively;
k_1 = a coefficient that accounts for the bond properties of the bonded reinforcement (0.8 for high bond bars);
k_2 = a coefficient that accounts for the distribution of strain (0.5 for bending and 1.0 for pure tension).

Earlier in Chapter 2 it was mentioned how, because of the potentially greater values of cover required by the new codes, the value of c above could result in greater percentages of steel required to control the crack width (as can been in Eq. (3.18) above, $S_{r,\,max}$ directly influences the crack width at the surface of the concrete as predicted by the code). Investigations suggest that the sides of the crack are in fact approximately parallel near the concrete surface but then at a point half the depth of the cover (depending on the cover dimension) the crack width reduces linearly to a value that is approximately 25% of the width at the surface (Forth and Beeby, 2012). It is therefore reasonable to ask whether the designer should actually be concerned with the crack width at the surface as predicted by the code and would it, in fact, be more reasonable to consider, for instance $c_{min,\,dur}$ only instead of c (i.e. c_{nom})? To recognise this issue, the UK NAD has adopted a simplified approach of reducing predicted surface crack widths; part of its aim is an attempt to reassure designers when specifying required / perceived excessive covers to suit durability requirements.

Kaethner (2011) questions whether BS EN 1992-1-1 has actually enhanced crack width prediction and among many interesting and valid points (including the effect of $\rho_{p,\,eff}$ and bar layout on crack width, although this is less critical for water-retaining structures where detailing tends to be more regular) she discusses the implications of excessive cover on predicted crack widths and their relevance for cases of flexural cracking, citing the work by Tammo and Thelandersson (2009). These researchers consider Equation (3.18) above split in terms of 'cover zone cracking' (k_3c) and 'cracking near bar' ($0.425\ k_1\ \phi\ /\ \rho_{p,\,eff}$) in an attempt to show that crack width is related to both cover and 'bond slip' theory. Their findings suggested that the 'cracking near bar' was not completely represented by the 'slip' theory and that cover also influenced at this position (this is further discussed in Chapter 5). Kaethner recommends a blanket reduction in the surface crack width prediction of 50 to 60% (this in effect represents the actual residual opening at the bar surface). Research by the authors confirms that for flexural elements, the value of k_3 should in fact be 2.1, which is 1.7 times less than the k_3 value of 3.4 (for axial members) or a 62% reduction (Kong *et al.*, 2007).

Expression 7.11 of BS EN 1992-1-1 represents the crack spacing predicted in the region over a bar–a distance of $5(c + \phi/2)$ with the bar centred in the middle of this distance. Where the final spacing of the bonded reinforcement exceeds $5(c + \phi/2)$ or where there is no bonded reinforcement within the tension zone, an upper bound to the crack width may be found by assuming a maximum crack spacing:

$$s_{r,max} = 1.3\ (h–x) \tag{3.19}$$

This approach by Part 3 introduces us to the second difference of note between the two codes.

The approaches in the two codes are similar; however, there are noteworthy differences. Firstly, whereas BS 8007 took account of the position of the crack in relation to the reinforcement bar (using a_{cr}–distance from the point considered to the surface of the nearest longitudinal bar), BS EN 1992-3 assumes an average crack width over a distance $5(c + \phi/2)$, this distance being centred around the bar centre-line. Of interest to designers is the maximum crack width, this is important in terms of corrosion and appearance etc., and so an average crack width, which is actually weighted towards the minimum crack width found directly over the bar (in BS 8007 terms, when $a_{cr} = c_{min}$; giving w_0) is possibly not the most appropriate value to consider. As mentioned above, BS EN 1992-1-1 does make reference to a region where the spacing of the bonded reinforcement exceeds $5(c + \phi/2)$ or where there is no bonded reinforcement within the tension zone and for this region it recommends an upper bound crack width by assuming a maximum crack spacing.

3.4.3 Comparison of Expression 7.9 (BS EN 1992-1-1) with Expression M1 (BS EN 1992-3)

The expression $(\varepsilon_{sm} - \varepsilon_{cm})$ (Eq. (3.17) above) represents the average strain in the steel (or the restrained component of strain [21]). However, in the two expressions (7.9 in BS EN 1992-1-1 and M1 in BS EN 1992-3), the average strain is derived from two different loading cases. In Expression 7.9, the average strain in the steel is the flexural strain in the steel at the crack, calculated on the basis of a cracked section, less the average strain in the concrete and less the tension stiffening provided by the concrete between the cracks. Whereas, in Expression M1, the average strain in the steel is equal to the pure tension strain in the steel, accounting for tension stiffening, less the average strain in the concrete at the surface (all resulting from the imposed deformations that arise from shrinkage and/or change in temperature). The two expressions are similar except for this important subtle difference. Expression M1 is derived in full along with a complete description of the principles of cracking in Chapter 5. The following breakdown of Expression 7.9 is intended to clarify the calculation of the strain in the steel and concrete and the quantification of the effects of tension stiffening.

For the FLEXURAL case (Expression 7.9):

$$(\varepsilon_{sm} - \varepsilon_{cm}) = [\sigma_s - k_t (f_{ct,eff} / \rho_{p,eff})(1 + \alpha_e \rho_{p,eff})] / E_s \quad (3.20)$$

Or

$$(\varepsilon_{sm} - \varepsilon_{cm}) = \sigma_s / E_s - [k_t (f_{ct,eff} / \rho_{p,eff})(1 + \alpha_e \rho_{p,eff})] / E_s \quad (3.21)$$

where σ_s / E_s = average maximum strain in the steel due to moment, calculated on the basis of a cracked section.

Further expansion of the second term in (3.21) gives:

$$[k_t (f_{ct,eff} / \rho_{p,eff})] / E_s + [(k_t (f_{ct,eff} / \rho_{p,eff}))(\alpha_e \rho_{p,eff})] / E_s \quad (3.22)$$

Reducing the second term in (3.22), using $\alpha_e = E_s/E_c$ gives:

$$k_t f_{ct,eff} / E_c$$

So (3.22) becomes:

$$[k_t (f_{ct,eff} / \rho_{p,eff})] / E_s + k_t f_{ct,eff} / E_c \quad (3.23)$$

where the second term of Eq. (3.23) equals the strain in the concrete between the cracks (depending on the duration of the load). By comparing the first term of (3.23) with Equation 13 of BS 8110 Part 2 or the theory presented by Beeby (1979) where tension stiffening correction at the level of the reinforcement is represented as:

$$\Delta\varepsilon = K [(ftf_{scr}) / (E_s \rho f_t)] \quad (3.24)$$

here f_{scr} (steel stress at cracking) = f_1 (steel stress under load considered), it is clear that the first term of (3.23) represents tension stiffening.

Therefore, for the FLEXURAL case, ε_{cm} is represented by the second term of Eq. (3.23) and ε_{sm} is represented by the first term of Eq. (3.21) (average maximum strain in the steel due to applied moment) less the first term of Eq. (3.23) (effect of tension stiffening provided by the concrete between the cracks). So the average strain in the steel is equivalent to the flexural strain in the steel at the crack calculated on the basis of a cracked section less the average strain in the concrete and less the tension stiffening provided by the concrete between the cracks.

Whereas, for a pure AXIAL tension case (Expression M1):

$$(\varepsilon_{sm} - \varepsilon_{cm}) = 0.5\alpha_e k_c k f_{ct,eff} (1 + 1/\alpha_e \rho) / E_s \quad (3.25)$$

where the stress in the steel, σ_s is:

$$\sigma_s = 0.5\alpha_e\, k_c k f_{ct,eff}\,(1 + 1/\alpha_e\rho) \tag{3.26}$$

The average strain in the steel $(\varepsilon_{sm} - \varepsilon_{cm})$ is purely due to the strain in the reinforcement (due to thermal or shrinkage strain), with tension stiffening accounted for (i.e. ε_{sm}) less the average strain in the concrete at the surface, ε_{cm}. (As can be seen in the derivation of Eq. (3.33) in Section 3.5, the tensile strength of the concrete and hence the tensile force carried by the concrete is considered. Defining tension stiffening as the uncracked regions between primary cracks which will help to stiffen the beam, clearly tension stiffening is represented in Eq. (3.26).)

3.5 Strength calculations

The analysis of the ultimate flexural strength of a section is made using formulae applicable to the design of normal structures. The partial safety factor for loads due to liquid pressure is taken as $\gamma_f = 1.2$. The formulae for the calculation of the ultimate limit state condition are obtained from a consideration of the forces of equilibrium and the shape of the concrete stress block at failure, and the following formulae are based on the recommendations of BS EN 1992-1-1.

The partial safety factor for concrete is taken as $\gamma_c = 1.5$ and for steel $\gamma_s = 1.15$. After allowing for the partial safety factor for concrete, for the UK practice of testing concrete strength using cubes, and for the equivalent rectangular stress block, a value of $0.57 f_{ck}$ is used for the width of the stress block, and a depth, s equal to $0.8 \times$ depth to the neutral axis (for concrete class ≤ C50/60).

Using the rectangular stress block as illustrated in Figure 3.8, the following equations may be derived:

$$\text{Lever arm factor } z_1 = 1 - 0.40x_1 \tag{3.27}$$

Force of tension = force of compression

Therefore,

$$A_s f_{yk} / 1.15 = 0.57 f_{ck} b 0.8 x_1 d \tag{3.28}$$

and

$$x_1 = 2.42 A_s f_{yk} / f_{ck}\, bd \tag{3.29}$$

Hence,

$$z_1 = 1 - 0.97 A_s f_{yk} / f_{ck} bd \tag{3.30}$$

Moment of resistance based on steel

$$M = 0.87 A_s f_{yk} z_1 d \tag{3.31}$$

With the maximum permissible value of $x = 0.45d$ (for concrete class ≤ C50/60 and ensuring a ductile section with gradual tension failure of the steel at the ultimate limit state), the maximum moment of resistance based on the concrete section is

$$M_u = 0.167 f_{ck} bd^2 \tag{3.32}$$

M_u represents the maximum ultimate moment that can be applied to the section without using compression reinforcement (a singly reinforced section). The actual applied ultimate moment M should be less than M_u.

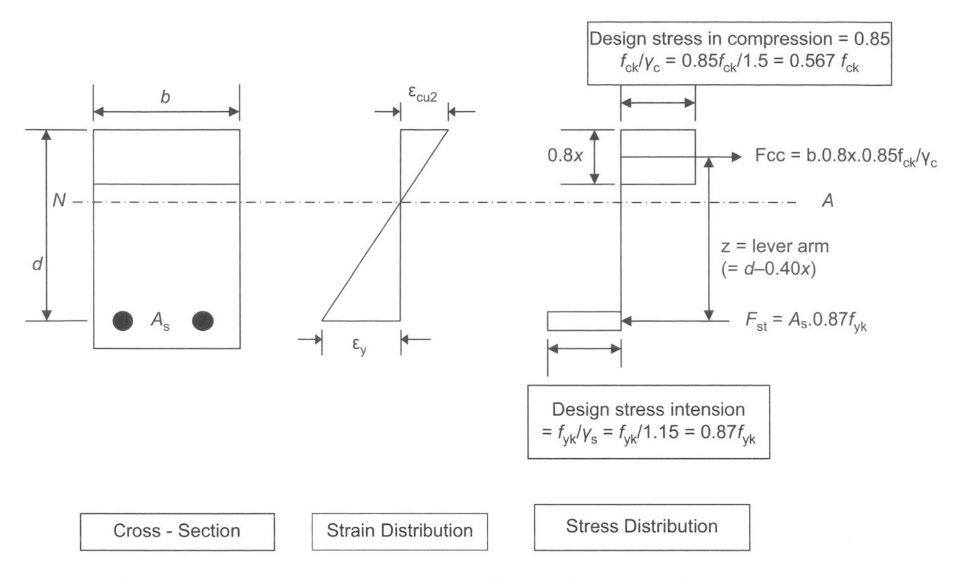

Figure 3.8 *Assumed stress diagrams – ultimate flexural limit state design.*

To calculate the area of reinforcement required to provide a given ultimate moment of resistance, it is convenient to rearrange Eqs (3.27) to (3.32) to provide the depth of the neutral axis in terms of the applied ultimate moment and the maximum ultimate moment. Rearrange (3.31) and (3.32) for $A_s f_{yk}$ and $f_{ck}bd$ in terms of M and M_u, respectively. Substitute into (3.28) and then rearrange for x_1. Replace z_1 using (3.27) and rearrange to produce a quadratic equation with x_1 as the unknown. Solving this quadratic equation gives (3.33) below:

$$x_1 = [1 - \sqrt{(1 - 0.59(M / M_u))}] / 0.8 \qquad (3.33)$$

This value may be substituted in Eqs (3.27) and (3.31) to calculate the required area of reinforcement.

After the arrangement of reinforcement has been decided, the ultimate shear stress should be rechecked (see Section 3.2.4).

3.6 Calculation of crack widths due to combined tension and bending (compression present)

3.6.1 Defining the problem

Some judgement is usually required when estimating crack widths due to the effects of direct tension combined with bending. The solution of the equations for a section under bending forces is straightforward as demonstrated in Section 3.4. The depth of the neutral axis can be calculated without difficulty. However, when tensile force is added to a section in bending, the position of the neutral axis changes so that a smaller fraction of the concrete section is in compression. Should more tension be applied, the neutral axis can subsequently move outside the section, and the whole section will then be in tension. This transition from pure flexure to pure tension is shown in Figure 3.9. Where minimum reinforcement is provided, BS EN 1992-3 provides guidance,

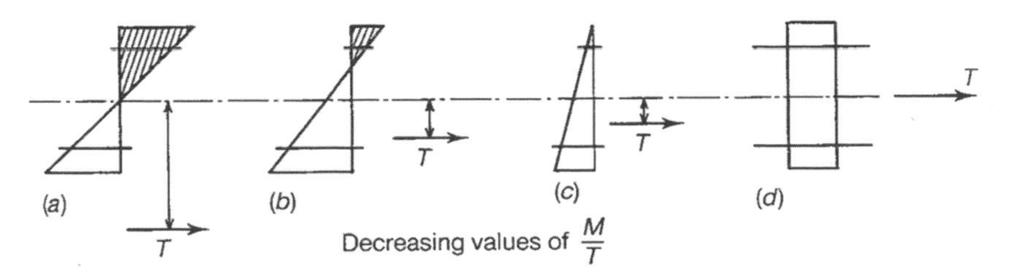

Figure 3.9 *Section with compressive stress.*

without direct calculation, on maximum bar diameters and bar spacings for various design crack widths for cases where the whole section is in tension. (This guidance supersedes that provided in BS EN 1992-1-1.) The tables provided to control axial tension cracking using these simplified detailing rules are based on the crack width formulae described in clause 7.3.4 of BS EN 1992-1-1. The obvious practical situation where pure tension will occur is in a cylindrical structure with a sliding joint at the base. Chapter 4 considers the design of a cylindrical tank.

When considering combined tension and bending, the most satisfactory approach is to consider the relation between the applied bending moment and the applied tensile force. The ratio M/T gives the value of the necessary eccentricity of a tensile force to produce the bending moment. A large value of M/T in relation to the section thickness indicates that the bending moment is predominant. A small value of M/T indicates that tension predominates (Figure 3.9). If one of the applied forces is small the simplest design approach is to prepare a design for the predominant force, and then to modify it by approximate methods.

Formulae for a section subject to applied tension and bending when both tensile and compressive stresses occur across the section can be derived from principles of strain compatibility and using the modular ratio method of elastic design. This approach for bending combined with tension produces formulae that are cubic in form and, therefore, a direct design is not possible. It applies when the tensile applied force is not too large. Initially, it is convenient to prepare a design for bending only and then to modify it by adding a modest amount of reinforcement. The equations may then be used to check the allowable values of applied loads. It is not possible to use the formulae without assuming a concrete section together with a quantity of reinforcement. Also, as there are three variables, an additional equation is needed. This is supplied by the crack width equation as it is the crack width that must ultimately be satisfied (see Example in the next section). The formulae for calculating the applied tensile force (N) and service bending moment (M) are derived below.

3.6.2 Formulae

The section geometry, strain distribution and the forces acting on a section in combined tension and flexure with the section in part compression are shown in Figure 3.9.

From strain compatibility (refer to Figure 3.10), the strain in the compression reinforcement ε_{sc} is given by

$$\varepsilon_{sc} = \varepsilon_{cc} \frac{x - a}{x} \tag{3.34}$$

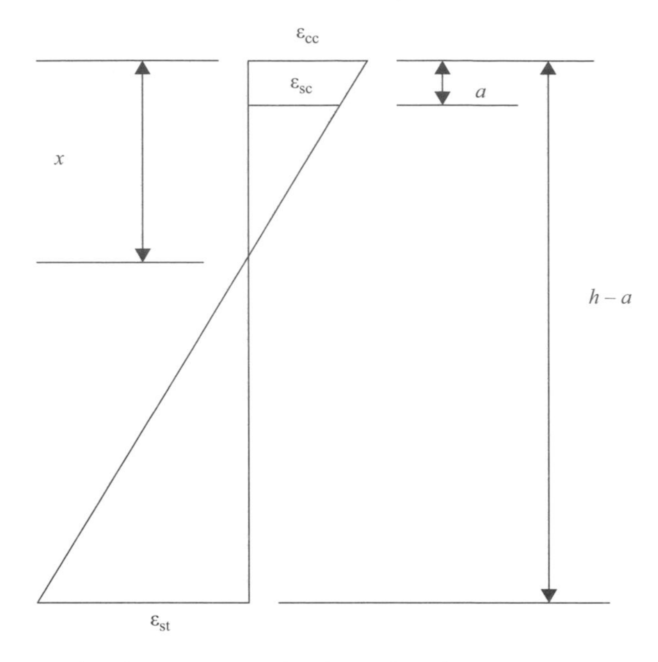

Figure 3.10 *Strain distribution across a doubly reinforced section.*

where

x is the depth to the neutral axis from the extreme compression fibre
a is the axis distance to the centroid of the reinforcement
ε_{cc} is the strain in the extreme compression fibre of the concrete

As both the reinforcement and concrete behaviour may be considered linear elastic, then the stress in the reinforcement σ_{sc} can be written as

$$\sigma_{sc} = E_s \varepsilon_{sc} \tag{3.35}$$

and the stress in the concrete σ_{cc} as

$$\sigma_{cc} = E_c \varepsilon_{cc} \tag{3.36}$$

where E_s and E_c are the elastic moduli of the steel and concrete, respectively. Thus σ_{sc} is given by

$$\sigma_{sc} = \frac{E_s}{E_c}\sigma_{cc}\left(1 - \frac{a}{x}\right) = \alpha_e \sigma_{cc}\left(1 - \frac{a}{h}\frac{h}{x}\right) \tag{3.37}$$

where α_e is the modular ratio.

The force F_s' in the compression steel is given by

$$F_s' = A_s' \sigma_{sc} = A_s' \alpha_e \sigma_{cc}\left(1 - \frac{a}{h}\frac{h}{x}\right) \tag{3.38}$$

where

A'_s is the area of the compression steel

Note, in deriving the compression reinforcement, the area of the concrete displaced by the steel has been ignored.

Again, from strain compatibility, the strain in the tension reinforcement ε_{st} is given by

$$\varepsilon_{st} = \varepsilon_{cc} \frac{h - x - a}{x} \tag{3.39}$$

where

x is the depth to the neutral axis from the extreme compression fibre
a is the axis distance to the centroid of the reinforcement
ε_{cc} is the strain in the extreme compression fibre of the concrete

As both the reinforcement and concrete behaviour may be considered linear elastic, then the stress in the reinforcement σ_{st} can be written as

$$\sigma_{st} = E_s \varepsilon_{st} \tag{3.40}$$

Thus σ_s is given by

$$\sigma_{st} = \frac{E_s}{E_c} \sigma_{cc} \left(\frac{h}{x} - \frac{a}{x} - 1 \right) = \alpha_e \sigma_{cc} \left(\frac{h}{x} - \frac{a}{h} \frac{h}{x} - 1 \right) \tag{3.41}$$

The force F_s in the tension steel is given by

$$F_s = A_s \sigma_{st} = A_s \alpha_e \sigma_{cc} \left(\frac{h}{x} - \frac{a}{h} \frac{h}{x} - 1 \right) \tag{3.42}$$

where

A_s is the area of the tension steel

The force in the concrete F_c is given by

$$F_c = \frac{b x \sigma_{cc}}{2} \tag{3.43}$$

From overall force equilibrium,

$$-N = F_c + F_s' - F_s \tag{3.44}$$

where N the axial tension force applied at the centroid of the section.

Rewriting Eq. (3.44) gives

$$-N = \frac{b x \sigma_{cc}}{2} + A_s' \alpha_e \sigma_{cc} \left(1 - \frac{a}{h} \frac{h}{x} \right) - A_s \alpha_e \sigma_{cc} \left(\frac{h}{x} - \frac{a}{h} \frac{h}{x} - 1 \right) \tag{3.45}$$

or

$$-\frac{N}{b h \sigma_{cc}} = \frac{1}{2} \frac{x}{h} + \frac{A_s'}{bh} \alpha_e \left(1 - \frac{a}{h} \frac{h}{x} \right) - \frac{A_s}{bh} \alpha_e \left(\frac{h}{x} - \frac{a}{h} \frac{h}{x} - 1 \right) \tag{3.46}$$

Defining a compression steel ratio ρ_c as

$$\rho_c = \frac{A_s'}{bh} \tag{3.47}$$

and a tension steel ratio ρ_t as

$$\rho_t = \frac{A_s}{bh} \tag{3.48}$$

then Eq. (3.46) may be rewritten as

$$-\frac{N}{bh\sigma_{cc}} = \frac{1}{2}\frac{x}{h} + \rho_c\alpha_e\left(1 - \frac{a}{h}\frac{h}{x}\right) - \rho_t\alpha_e\left(\frac{h}{x} - \frac{a}{h}\frac{h}{x} - 1\right) \tag{3.49}$$

Taking moments about the centre-line (or mid-depth) of the section,

$$M = F_c\left[\frac{h}{2} - \frac{x}{3}\right] + \left[F_s' + F_s\right]\left[\frac{h - 2a}{2}\right] \tag{3.50}$$

or

$$M = \frac{bx\sigma_{cc}}{2}\left[\frac{h}{2} - \frac{x}{3}\right] + \left[A_s'\alpha_e\sigma_{cc}\left(1 - \frac{a}{h}\frac{h}{x}\right) + A_s\alpha_e\sigma_{cc}\left(\frac{h}{x} - \frac{a}{h}\frac{h}{x} - 1\right)\right]\left[\frac{h - 2a}{2}\right] \tag{3.51}$$

or

$$\frac{M}{bh^2\sigma_{cc}} = \frac{1}{2}\frac{x}{h}\left[\frac{1}{2} - \frac{1}{3}\frac{x}{h}\right] + \left[\rho_c\alpha_e\left(1 - \frac{a}{h}\frac{h}{x}\right) + \rho_t\alpha_e\left(\frac{h}{x} - \frac{a}{h}\frac{h}{x} - 1\right)\right]\left[\frac{1}{2} - \frac{a}{h}\right] \tag{3.52}$$

The same reinforcement will be needed in each face if the moment is reversible. If the moment can only act in one direction it may be possible to design on the basis of differing amounts of reinforcement in each face.

Even if, as is usual, the tension reinforcement and the compression reinforcement are set equal, i.e.

$$\rho_c = \rho_t = \rho \tag{3.53}$$

Eqs (3.49) and (3.52) cannot be solved as there are three variables. An additional equation is therefore needed. This is supplied by the crack width equation as it is the design crack width that must be satisfied. The crack width is dependent upon the crack spacing and the mean strain (allowing for tension stiffening) in the reinforcement. Thus it is readily apparent that a closed form solution for a given design is not possible. This means an iterative method must be adopted, for example the use of a simple spreadsheet.

The simplest manner of proceeding is:

- Estimate h (if not already known) and a. (It is easier to work in terms of the axis distance although it should be checked that in a given case the cover is acceptable.)
- Estimate the bar size and spacing.
- Estimate x and calculate $\frac{x}{h}$ (Note $0.5 > \frac{x}{h} > \frac{a}{h}$

- Determine σ_{cc} from Eq. (3.49).
- Using the value of σ_{cc}, determine M from Eq. (3.52). If the value of M is less than the applied moment, then increase the amount of reinforcement (or section size, if appropriate). If M is greater than the applied moment, reduce the amount of reinforcement (or section size) until M is only marginally greater than the applied moment.
- The stress in the tension reinforcement may then be determined from Eq. (3.41) and the crack width calculated.
- If the crack width is unsatisfactory, it will be necessary to iterate through the complete calculations.

Example 3.2

Determine the reinforcement required (assumed equal in each face) for a section 300 mm thick carrying actions of a tensile force of 78 kN/m and a moment of 57 kNm/m at serviceability limit state assuming an axis distance of 50 mm to the centroid of the reinforcement and an allowable crack width of 0.2 mm.

Take the value of α_e as 15 and assume the loading is long term. The value of b is taken as 1 000 mm. Assume C30/37 concrete and that the value of $f_{ct,eff}$ may be taken as f_{ctm}.

The calculations given below are the final iteration only.

From Table 3.1 (BS EN 1992-1-1) determine f_{ctm}:

$$f_{ctm} = 0.30 f_{ck}^{2/3} = 0.30 \times 30^{2/3} = 2.9\text{MPa}$$

Assume reinforcement of B20 at 200 mm centres (A_s = 1570 mm²/m)

$$\rho = \frac{A_s}{bh} = \frac{1570}{1000 \times 300} = 0.00524$$

$$\alpha_e \rho = 15 \times 0.00524 = 0.0786$$

For a value of x = 72 mm, or $\frac{x}{h}$ of 0.24, determine $\frac{N}{bh\sigma_{cc}}$ from Eq. (3.49)

$$\frac{1}{2}\frac{x}{h} + \rho_c \alpha_e \left(1 - \frac{a}{h}\frac{h}{x}\right) - \rho_t \alpha_e \left(\frac{h}{x} - \frac{a}{h}\frac{h}{x} - 1\right) =$$

$$\frac{1}{2} \times 0.24 + 0.0786\left(1 - \frac{0.167}{0.24}\right) - 0.0786\left(\frac{1}{0.24} - \frac{0.167}{0.24} - 1\right) = -0.0503$$

Hence,

$$\sigma_{cc} = \frac{-78 \times 10^3}{-0.0503 \times 1000 \times 300} = 5.17 \text{ MPa}$$

Determine $\frac{M}{bh^2\sigma_{cc}}$ from Eq. (3.52)

$$\frac{M}{bh^2\sigma_{cc}} = \frac{1}{2}\frac{x}{h}\left[\frac{1}{2} - \frac{1}{3}\frac{x}{h}\right] + \left[\rho_c\alpha_e\left(1 - \frac{a}{h}\frac{h}{x}\right) + \rho_t\alpha_e\left(\frac{h}{x} - \frac{a}{h}\frac{h}{x} - 1\right)\right]\left[\frac{1}{2} - \frac{a}{h}\right]$$

$$= \frac{1}{2}\times 0.24\left[\frac{1}{2} - \frac{1}{3}\times 0.24\right]$$

$$+0.0786\left[\left(1 - \frac{0.167}{0.24}\right) + \left(\frac{1}{0.24} - \frac{0.167}{0.24} - 1\right)\right]\left[\frac{1}{2} - 0.167\right] = 0.123$$

$$M = 0.123bh^2\sigma_{cc} = 0.123\times 1000\times 300^2\times 5.17\times 10^{-6} = 57.2\text{kNm/m}$$

This is satisfactory as far as equilibrium is concerned (applied moment is 57 kNm/m).

Determine σ_{st} from Eq. (3.41)

$$\sigma_{st} = \alpha_e\sigma_{cc}\left(\frac{h}{x} - \frac{a}{h}\frac{h}{x} - 1\right) = 15\times 5.17\left(\frac{1}{0.24} - \frac{0.167}{0.24} - 1\right) = 192\text{MPa}$$

Limiting spacing of reinforcement is $5(c + \phi/2)$:

$$5\left(c + \frac{\varphi}{2}\right) = 5\left(a - \frac{\varphi}{2} + \frac{\varphi}{2}\right) = 5a = 5\times 50 = 250 \text{ mm}$$

Actual spacing is 200 mm, therefore use Eq (3.18) to determine $s_{r,max}$:

$k_1 = 0.8$ (high bond bars)
$k_2 = 0.5$ (flexure)
$k_3 = 3.4$ (UK National Annexe to BS EN 1992-1-1)
$k_4 = 0.425$ (UK National Annexe to BS EN 1992-1-1)

The cover c is given by

$$c = a - \frac{\varphi}{2} = 50 - \frac{20}{2} = 40\,mm$$

Determine $A_{c,eff}$:

$A_{c,eff}$ is the minimum of $2.5(h - d)b$, $(h - x)b/3$ and $hb/2$.

$$2.5(h - d)b = 2.5(300 - (300 - 50))\times 1000 = 125000 \text{ mm}^2$$

$$\frac{(h - x)b}{3} = \frac{(300 - 72)1000}{3} = 76000 \text{ mm}^2$$

$$\frac{hb}{2} = \frac{300\times 1000}{2} = 150000 \text{ mm}^2$$

The least value is 76 000 mm², thus $A_{c,eff}$= 76 000 mm².

The effective reinforcement ratio $\rho_{p,eff}$ (in the absence of prestressing tendons) is given by

$$\rho_{p,eff} = \frac{A_s}{A_{c,eff}} = \frac{1570}{76000} = 0.0207$$

$$S_{r,max} = k_3 c + k_1 k_2 k_4 \frac{\phi}{\rho_{p,eff}} = 3.4 \times 40 + 0.8 \times 0.5 \times 0.425 \frac{20}{0.0207} = 300 \text{ mm}$$

Determine $\varepsilon_{sm} - \varepsilon_{cm}$

$$\varepsilon_{sm} - \varepsilon_{cm} = \frac{\sigma_s - k_1 \frac{f_{ct,eff}}{\rho_{p.eff}}\left(1 + \alpha_e \rho_{p.eff}\right)}{E_s} \geq 0.6 \frac{\sigma_s}{E_s}$$

The parameter k_1 is the load duration factor (= 0.4 for long-term load). It is not to be confused with the k_1 factor in the crack spacing formula.

Also, σ_s is the stress in the tension reinforcement assuming a cracked section, which is the same as σ_{st} in Eq. (3.41).

$$\frac{\sigma_s - k_1 \frac{f_{ct,eff}}{\rho_{p,eff}}\left(1 + \alpha_e \rho_{p,eff}\right)}{E_s} = \frac{192 - 0.4 \frac{2.9}{0.0207}\left(1 + 15 \times 0.0207\right)}{200 \times 10^3} = 593 \times 10^{-6}$$

$$0.6 \frac{\sigma_s}{E_s} = 0.6 \frac{192}{200 \times 10^3} = 576 \times 10^{-6}$$

The value of $\varepsilon_{sm} - \varepsilon_{cm}$ to be used to calculate the crack width is therefore 593×10^{-6}.

From Eq (3.16), the characteristic crack width w_k is given by

$$w_k = s_{r,max}\left(\varepsilon_{sm} - \varepsilon_{cm}\right) = 300 \times 593 \times 10^{-6} = 0.178 \text{ mm}$$

Note, within practical bar spacings it is not possible to get closer to the allowable limit of 0.2 mm.

As an exercise it was decided to investigate the effect of differing amounts of reinforcement in each face. The tension reinforcement was retained as B20 at 200 mm centres (1570 mm²/m), but the compression was approximately halved to B16 at 250 mm centres (804 mm²/m).

The calculations gave a depth of compression zone in the concrete of 74.4 mm, a resultant moment capacity of 57.3 kNm/m, a concrete stress of 5.4 MPa, a tensile steel stress of 191 MPa and a final crack width of 0.177 mm. Thus in this particular case there could be an economy made in reducing the reinforcement in the compression face. This would only be possible if the applied moment could only act in one direction.

3.7 Detailing

The reinforcement detailing requirements for water-retaining structures follow the usual rules for normal structures. Guidance can be found in Sections 7, 8 and 9 of EC2 Part 1 with additional information available in BS EN 1992-3, mainly relating to prestressed members. Primarily, the code is concerned with controlling cracking and therefore promoting adequate bond via sufficient concrete cover, satisfactory placement and compaction of concrete etc., and transverse reinforcement to deal with the potential development of tensile stresses in high bond stress regions. Bars should be detailed for continuity on the liquid faces and sudden changes of reinforcement ratio should be avoided. The distribution reinforcement in walls should be placed in the outer layers if its maximum effect is to be achieved. Spacers should be detailed to ensure that the correct cover is maintained (Figure 3.11).

3.7.1 Spacing and bar diameter

Guidance is provided on the minimum clear distance between horizontal or vertical bars or between layers of parallel bars. Three minimum limits to spacing are provided: 20 mm, maximum aggregate size plus 5 mm (k_2 in the NAD), and bar diameter × k_1 (where k_1 = 1 in the UK NAD).

For slabs with an overall thickness of greater than 200 mm, Cl 7.3.3 of EC2 controls cracking without direct calculation for specified crack widths of either 0.2 mm, 0.3 mm or 0.4 mm depending on the specific requirements of the structure. With the crack widths fixed, the rules for determining crack widths are presented in tabular form in BS EN 1992-1-1 for bar size (Table 7.2N) and bar spacing (Table 7.3N) for a range of steel stresses from 160 to 450 MPa. For cases where cracks are caused mainly by loading, bar diameter or bar spacing tables can be used; for cracking caused predominantly by restraint of imposed strain, only the bar size table can be utilised. The maximum bar diameters obtained from Table 7.2N are then modified by Eq. 7.6N for cases where bending forces dominate (i.e. at least part of the section is in compression) and Eq. 7.7N where the section is subject to uniform axial tension.

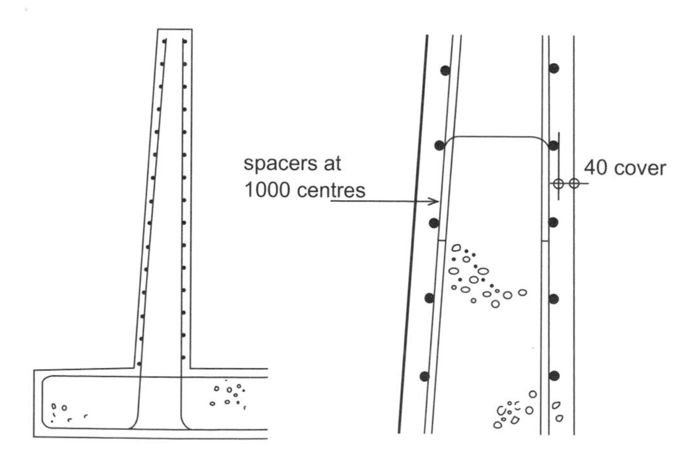

Figure 3.11 *Detailing of spacer reinforcement.*

Table 3.4 *Maximum bar diameter.*

Steel stress (MPa)	*BS EN 1992-1-1*	*Narayanan and Beeby*	*BS EN 1992-3*
160	32	32	-
200	25	25	35
240	16	20	22
280	12	16	16
320	10	12	12
360	8	10	9
400	6	8	6
450	5	-	5

BS EN 1992-3 further modifies Eq. 7.7N to reflect the fact that the basic guidance in BS EN 1992-1-1 has been obtained from studies of elements subjected to pure flexure. The expression BS EN 1992-3 provides (7.122) should in theory require smaller maximum bar diameters (at closer centres) and it presents the bar diameter and spacings data not in tabular form but in graphical form (Figure 7.103N and 7.104N, respectively).

Tables 3.4 and 3.5 compare the maximum bar diameter and spacing, respectively, of tension bars for a crack width of 0.3 mm as suggested by BS EN 1992-1-1 and BS EN 1992-3 and Narayanan and Beeby (2005).

Some of the data contained within Tables 3.4 and 3.5 are extracted from figures [1, 2, 49], and as such there may be minor errors in the values presented. The values from BS EN 1992-1-1 are for the case of bending only. The bar diameter values taken from Narayanan and Beeby are for bending only; however, the maximum bar spacing values are for both bending and pure tension (pure tension values are shown in brackets). All values taken from BS EN 1992-3 are for pure tension. Interestingly it suggests bigger bar diameters than BS EN 1992-1-1 except at high steel stresses; this contradicts the purpose of Expression 7.122, which suggests the use of smaller bar sizes. The bar spacings recommended in BS EN 1992-3, which are for pure tension cases, are mostly less than those recommended in BS EN 1992-1-1, which are for flexural cases. However, the values suggested by BS EN 1992-3 are in all cases greater than the values recommended by Narayanan and Beeby. For flexural cases, the area of concrete in tension, just before formation of the first crack, A_{ct} will be approximately half the overall depth of the section, whereas, for an element in pure tension, all of the section will be in tension, i.e. A_{ct} will be based on a tension zone with depth equal to the overall depth of the section. Also, for pure flexure, k_c (the stress distribution constant) is equal to 0.4, whereas for pure tension, $k_c = 1.0$. The recommendation by Narayanan and Beeby that the maximum bar spacing for cases of pure tension should be 50% of those spaces recommended for flexural cases therefore seems reasonable. Comparisons can also be drawn for a fixed crack width of 0.2 mm. The values presented in BS EN 1992-3 are more reasonable, except at low stresses.

Table 3.5 *Maximum bar spacing. Narayanan and Beeby (2005).*

Steel stress (MPa)	*BS EN 1992-1-1*	*Narayanan and Beeby*	*BS EN 1992-3*
160	300	300 (150)	-
200	250	250 (125)	260
240	200	200 (100)	160
280	150	150 (75)	90
320	100	100 (50)	60
360	50	50 (25)	36
400	-	-	20

3.7.2 Anchorage and laps

Clear guidance on anchorage of reinforcement and laps in reinforcement can be found in BS EN 1992-1-1: Sections 8 and 9.Very little additional guidance is provided in BS EN 1992-3. In summary, bars must be sufficiently anchored so as to develop their design stresses. The bond between the bar and the concrete dictates the anchorage length; guidance is provided on whether the bond conditions are good or poor–this relates to the angle of inclination of the bar and the depth of concrete (in the direction in which the concrete is poured) in which the bar is located. The design anchorage length, l_{bd} is obtained by modifying the basic required anchorage length, $l_{b,red}$ using five α factors. These factors account for the shape of the bar, the concrete cover, the confinement by transverse reinforcement (not welded or welded to the main reinforcement) and confinement by transverse pressure. Where lapping is required to transfer stresses from one bar to another, the laps should be staggered and not in regions of high stress. The length of a lap is based on the minimum anchorage length, which as with the design anchorage length, is modified using the five α factors listed above plus an additional α factor, which accounts for the percentage of lapped bars relative to the total cross-section area.

Chapter 4
Design of Prestressed Concrete

by Dr John A. Purkiss, Consultant

In prestressed concrete the material is put under a state of compressive stress before the imposition of imposed actions. The magnitude of the induced stresses is such that after the imposition of the applied actions the stresses in the concrete are still effectively compressive. Prestress can be applied in either or both of two orthogonal directions within the plane of a structural element. Prestressed concrete may be divided into two categories:

(i) Pre-tensioned: This is a factory-based method in which prestressing wires are tensioned on a bed; the concrete elements are then cast in steel moulds around the wires. After a sufficient gain in strength often using accelerated curing techniques, the bed is unstressed and the wires between the elements are then cut. This technique is more appropriate for precast floor units or bridge beams.

(ii) Post-tensioned: In the case of internal post-tensioning, ducts are cast into the concrete, tendons threaded through and then stressed on site. For external post-tensioning, tendons or wires are wrapped externally round the structure and then prestressed.

For liquid-retaining structures, prestressing would appear to possess considerable advantage over reinforced concrete as the member remains in compression with no cracks and therefore no leakage. For liquid-retaining structures, post-tensioning is more appropriate, but it should be noted that there are disadvantages over reinforced concrete:

- Cost: This is in part due to the procedure itself and part due to the necessity of employing specialist contractors.
- Safety: There are additional safety issues that need to be addressed, i.e. to allow for the potential hazards should an anchorage come adrift during the prestressing operation.

4.1 Materials

4.1.1 Concrete

As the maximum stress in the concrete will not be particularly high in relation to the strength of the concrete, a concrete strength as low as C45/55 may be employed. However, a relatively low strength concrete will lead to potentially high prestressing losses owing to the relatively low value of the concrete elastic modulus. With any strength concrete especially where the water cement ratio is low, adequate workability must be ensured using plasticisers or super-plasticisers.

4.1.2 Prestressing tendons

Generally low relaxation tendons are used (currently specified to BS 5896, which is being revised, although BS EN 10138-1 is in preparation and should eventually replace the British Standard) with strengths between 1 000 and 2 000 MPa. The jacking force P_{max} shall not exceed

$$P_{max} = A_p \sigma_{p,max} \tag{4.1}$$

where A_p is the area of the tendons and $\sigma_{p,max}$ is given by

$$\sigma_{p,max} = \text{minimum}(0.8 f_{pk}, 0.9 f_{p0.1k}) \tag{4.2}$$

where f_{pk} is the characteristic tensile strength and $f_{p0.1k}$ is the characteristic 0.01% proof stress (BS EN 1992-1-1, Cl 3.3.6(7)).

Note that a lower value of prestress might be appropriate, except that the prestress losses will then be more deleterious. UK practice has conventionally used an initial prestress level of 75% (Anchor, 1992).

4.1.3 Prestress losses

Prestress losses, generally expressed as a fractional (or percentage) value of the initial prestress, are caused by a number of factors:

- Relaxation of the steel: This is a form of creep under constant strain (i.e. due to creep the stress reduces; however, the overall strain in the steel remains the same) and is caused by changes in the microscopic structure of the steel, and generally expressed as a value at 1 000 hours. The value used in any particular design is taken from manufacturers' data for the particular prestressing tendon or cable in use.
- Elastic deformation of the concrete: As the concrete is loaded by the prestress it suffers a compressive strain, which shortens the length of the tendon and thereby reduces the prestress. For pretensioned bonded tendons this loss is unavoidable as the lengths of the tendon and the concrete member are identical. For post-tensioned concrete where on stressing the tendon is not bonded to the concrete, the losses can be counteracted by increasing the prestressing force, thereby ameliorating the losses.
- Shrinkage of the concrete: The shrinkage in concrete is a function of aggregate, water-cement ratio, age at transfer, section thickness and ambient relative humidity. These factors make exact determination of the values used for shrinkage strains difficult, although guidance is given in BS EN 1992-1-1.
- Creep of the concrete: Creep in concrete is a function of aggregate, water-cement ratio, age at transfer, section thickness, ambient relative humidity and ambient temperature. These factors make exact determination of the values

used for creep strains difficult, although guidance is given in BS EN 1992-1-1 through the use of a creep coefficient, φ.

- Anchorage slip: There will be some slip at anchorages when the anchorages bed in due to the application of the prestress force. In pre-tensioned concrete using a prestressing bed where anchorage slip can only occur at the ends of the bed, the value is small and may be allowed for by increasing the prestress force. For post-tensioned concrete, the slip must be allowed for. The value is generally known for a particular prestressing system and is more critical for shorter members.
- Friction loss: For pre-tensioned members this is not critical, but for post-tensioned members the tendons are within prestressing ducts and any movement during the stressing operation will cause frictional losses. For curved ducts this is more critical and must be allowed for.

Prestress loss must be determined for both at transfer and at full serviceability limit state. Typical values at full serviceability limit state are: relaxation (1–12%), elastic deformation (1–10%), shrinkage (1–6%), creep (5–15%), anchorage slip (0–5%) and friction (3–7%). At transfer and serviceability relaxation, anchorage slip and friction must be taken into account (Martin and Purkiss, 2006). Thus at full serviceability limit state assuming low relaxation tendons the losses could be as high as around 45% and at transfer 15% (note these values have been rounded up from the values given above).

There are two methods of determining the prestress losses. Either each loss may be considered in turn and values added, or a composite equation given in BS EN 1992-1-1 may be used. The theory presented below has been taken from Martin and Purkiss (2006). It should be noted that BS EN 1992-1-1 tends to work in terms of loss in prestressing force; when these values are expressed as percentage losses then they may equally be determined in terms of stresses.

Note, all the theory uses the conventional sign convention that compression is positive.

Relaxation loss

This has been covered above.

Elastic deformation

For bonded pretensioned concrete the elastic deformation may be calculated as follows:

If f_s is the loss of stress in the steel and σ_{cp} is the stress in the concrete adjacent to the steel at transfer, then for strain compatibility

$$\frac{\sigma_{cp}}{E_{cm}} = \frac{f_s}{E_s} \tag{4.3}$$

If f_{pi} is the initial prestress in the steel, then the percentage loss Δf_s is given by

$$\Delta f_s = 100\frac{f_s}{f_{pi}} = 100\frac{\sigma_{cp}}{f_{pi}}\frac{E_s}{E_{cm}} \tag{4.4}$$

The elastic modulus of prestressing wire or bar may be taken as 205 GPa and strand as 195 GPa (BS EN 1992-1-1, Cl 3.3.6(3)), although manufacturers' data may also be adopted.

The elastic modulus for concrete E_{cm} is given by

$$E_{cm} = 22\left(\frac{f_{ck}+8}{10}\right)^{0.3} \quad (4.5)$$

where f_{ck} is the concrete cylinder strength at transfer.

For post-tensioned members, either the elastic losses may be taken into account by increasing the prestress or they may be taken as half the value given by Eq. (4.4) (BS EN 1992-1-1, Cl 5.10.5.1).

Shrinkage (see also Chapter 3)

In the absence of specific experimental data, values of shrinkage strain may be taken from Table 3.2 of BS EN 1992-1-1 or for drying shrinkage from Annex B of the same document.

The total shrinkage strain ε_{cs} has two components–autogenous shrinkage ε_{ca} and drying shrinkage ε_{cd}.

Autogenous shrinkage

This develops in the early days after casting during hardening of the concrete. It needs considering when new concrete is cast against hardened concrete. From Cl 3.1.4 (BS EN 1992-1-1) the autogenous shrinkage strain at time t is entirely a function of the concrete strength and is given by

$$\varepsilon_{ca}(t) = \left[2.5\left(f_{ck}-10\right)\times 10^{-6}\right]\left[1-\exp\left(-0.2t^{0.5}\right)\right] \quad (4.6)$$

Drying shrinkage

The drying shrinkage ε_{cd} is given by Annex B (BS EN 1992-1-1) as

$$\varepsilon_{cd}(t) = \beta_{ds}(t,t_s)k_h\varepsilon_{cd,0} \quad (4.7)$$

where

$$\varepsilon_{cd,0} = 0.85\left[\left(220+110\alpha_{ds1}\right)\exp\left(-\alpha_{ds2}\frac{f_{cm}}{f_{cm0}}\right)\right]\beta_{RH}\times 10^{-6} \quad (4.8)$$

where

$$\beta_{RH} = 1.55\left[1-\left(\frac{RH}{RH_0}\right)^3\right] \quad (4.9)$$

and

f_{cm} is the mean compressive strength

f_{cm0} is a reference strength taken as 10 MPa

RH is the ambient relative humidity (%)

RH_0 is a reference value of 100% with α_{ds1} and α_{ds2} as coefficients dependent upon cement type. Values of α_{ds1} and α_{ds2} are given in Table 4.1 (from Cl B.2), together with values of s from Cl 3.1.2 (6) both from BS EN 1992-1-1

Table 4.1 *Values of α_{ds1}, α_{ds2} and s.*

Cement Class	α_{ds1}	α_{ds2}	s
S	3	0.13	0.20
N (Ordinary early strength)	4	0.12	0.25
R (High early strength)	5	0.11	0.38

The parameter $\beta_{ds}(t,t_s)$ is given by

$$\beta_{ds}((t,t_s) = \frac{t - t_s}{t - t_s + 0.04\sqrt{h_0^3}} \tag{4.10}$$

Note as $t \rightarrow \infty$, $\beta_{ds}(t,t_s) = 1.0$.
Values of k_h (Table 3.3, BS EN 1992-1-1) are given by
for $100 < h_0 < 200$ mm

$$k_h = 1 - 0.0015(h_0 - 100) \tag{4.11}$$

for $200 < h_0 < 300$ mm

$$k_h = 0.85 - 0.001(h_0 - 200) \tag{4.12}$$

for $300 < h_0 < 500$ mm

$$k_h = 0.75 - 0.00025(h_0 - 300) \tag{4.13}$$

for $h_0 > 500$ mm

$$k_h = 0.7 \tag{4.14}$$

where the parameter h_0 denotes a notional size of the member and is defined by

$$h_0 = \frac{2A_c}{u} \tag{4.15}$$

where A_c is the area of the concrete section and u is that part of the perimeter of the section exposed to drying. For a wall drying on two faces and the top, where $h >> t$, $h_0 \approx t$, where h is the height and t the thickness.

For external structures, Bamforth (2007) recommends a value of 85% RH and for internal structures a value of 45% RH.
The total shrinkage, ε_{cs} is given by

$$\varepsilon_{cs} = \varepsilon_{cd} + \varepsilon_{ca} \tag{4.16}$$

Assuming no slip between the tendons and the concrete, the strain in the tendons is then $\varepsilon_{ccl.0}$ and the percentage loss in prestress Δf_s is given by

$$\Delta f_s = 100\frac{f_s}{f_{pi}} = 100\frac{E_s\varepsilon_{cs}}{f_{pi}} \tag{4.17}$$

For both shrinkage and creep the relative humidity should be assessed on the levels prevalent at the early age of the structure before filling with liquid as both these phenomena are time dependent, with the majority of shrinkage or creep occurring very early in the life of the structure.

Creep

The creep strain $\varepsilon_{cc}(\infty,t_0)$ is given by

$$\varepsilon_{cc}(\infty,t_0) = \varphi(\infty,t_0)\frac{\sigma_c}{E_{cm}} \tag{4.18}$$

where

$\varphi(\infty,t_0)$ is the final creep coefficient
σ_c is the stress in the concrete
E_{cm} is the elastic modulus of the concrete

Assuming no slip between the tendons and the concrete, from strain compatibility the strain in the tendons is also equal to ε_{cs} and the percentage loss in prestress Δf_s is given by

$$\Delta f_s = 100\frac{f_s}{f_{pi}} = 100\varphi(\infty,t_0)\frac{\sigma_c}{f_{pi}}\frac{E_s}{E_{cm}} \tag{4.19}$$

Values of $\varphi(\infty,t_0)$ may be obtained from either Fig. 3.1 or Annex B, both from BS EN 1992-1-1. The equations from Annex B are given below but with the value of *t*, the time at which the creep coefficient is needed, set equal to infinity.

The creep coefficient $\varphi(\infty,t_0)$ is given by

$$\varphi(\infty,t_0) = \varphi_0\beta(\infty,t_0) \tag{4.20}$$

where φ_0 is given by

$$\varphi_0 = \varphi_{RH}\beta(f_{cm})\beta(t_0) \tag{4.21}$$

where t_0 is the age at loading (or prestressing).

The coefficient φ_{RH} allows for the effect of the ambient relative humidity *RH* (in per cent) on the creep coefficient and is given by
$f_{cm} \leq 35$ (MPa)

$$\varphi_{RH} = 1 + \frac{1-0.01(RH)}{0.1\sqrt[3]{h_0}} \tag{4.22}$$

$f_{cm} > 35$ (MPa)

$$\varphi_{RH} = \left[1 + \frac{1-0.01(RH)}{0.1\sqrt[3]{h_0}}\left[\frac{35}{f_{cm}}\right]^{0.7}\right]\left[\frac{35}{f_{cm}}\right]^{0.2} \tag{4.23}$$

The parameter h_0 is defined in Eq. (4.15).

The coefficient $\beta(f_{cm})$ allows for the effect of concrete strength on the notional creep coefficient and is given by

$$\beta(f_{cm}) = \frac{16.8}{\sqrt{f_{cm}}} \tag{4.24}$$

where f_{cm} is the concrete strength at 28 days.

The parameter $\beta(t_0)$ allows for the age of the concrete at loading (or prestressing) and is given by

$$\beta(t_0) = \frac{1}{0.1 + t_0^{0.20}} \tag{4.25}$$

The parameter $\beta_c(t,t_0)$ allows for the effect of development of creep after loading and is given by

$$\beta_c(t,t_0) = \left[\frac{t - t_0}{\beta_H + t - t_0}\right]^{0.3} \tag{4.26}$$

when $t \to \infty$, $\beta_c(t,t_0) = 1.0$.

For completeness the value of β_H is given by

$f_{cm} \le 35$ (MPa)

$$\beta_H = 1.5\left[1 + (0.012RH)^{18}\right]h_0 + 250 \le 1500 \tag{4.27}$$

$f_{cm} > 35$ (MPa)

$$\beta_H = 1.5\left[1 + (0.012RH)^{18}\right]h_0 + 250\left[\frac{35}{f_{cm}}\right]^{0.5} \le 1500\left[\frac{35}{f_{cm}}\right]^{0.5} \tag{4.28}$$

Anchorage slip

If the slip at the anchorage is δL and the length of the tendon is L, then the loss in prestress Δf_s is given by

$$\Delta f_s = 100\frac{\delta L}{L}\frac{E_s}{f_{p1}} \tag{4.29}$$

For a given prestressing system, the slip at the anchorage will be known or may be allowed for by increasing the prestress force.

Friction loss

For a straight duct, the percentage loss in prestress Δf_s is given by

$$\Delta f_s = 100\left(1 - e^{-\mu kx}\right) \tag{4.30}$$

where x is the distance from the prestressing jack, k is a parameter generally lying between 0.005 to 0.01 and μ is the coefficient of friction. Typical values of μ for internal tendons may be taken from Table 4.2 (Section 12.3.7, Martin and Purkiss, 2006).

If the duct is curved, then there are additional losses due to the curvature, which are given by

$$\Delta f_s = 100\left(1 - e^{-\mu\theta}\right) \tag{4.31}$$

Table 4.2 *Coefficient of friction μ for internal tendons.*

Type of tendon	*Coefficient of friction μ*
Cold drawn wire	0.17
Strand	0.19
Deformed bar	0.65
Smooth round bar	0.33

where θ is the sum of the absolute values of the angular displacements (irrespective of sign) from the jacking point.

The overall effect due to friction and curvature on the loss of prestress is therefore given by

$$\Delta f_{s} = 100\left(1 - e^{-\mu(\theta + kx)}\right) \tag{4.32}$$

4.1.4 Overall prediction of prestress loss ΔP_{c+s+f}

This is given by Eq. (5.46) in BS EN 1992-1-1 as

$$\Delta P_{c+s+f} = A_{p}\sigma_{p,c+s+f} = A_{p}\frac{\varepsilon_{cs}E_{p} + 0.8\Delta\sigma_{pr} + \dfrac{E_{p}}{E_{cm}}\varphi(t,t_{0})\sigma_{c,Qp}}{1 + \dfrac{E_{p}}{E_{cm}}\dfrac{A_{p}}{A_{c}}\left(1 + \dfrac{A_{c}}{I_{c}}z_{cp}^{2}\right)\left[1 + 0.8\varphi(t,t_{0})\right]} \tag{4.33}$$

where

$\Delta\sigma_{p,c+s+f}$ is the absolute value of the variation of stress in the tendons due to creep shrinkage and relaxation at location x and time t
ε_{cs} is the estimated shrinkage strain
E_{p} is the modulus of elasticity of the prestressing steel
E_{cm} is the elastic modulus of the concrete
$\Delta\sigma_{pr}$ is the variation in stress due to relaxation under a stress due to the initial prestress and the quasi-permanent actions (if appropriate)
$\varphi(t,t_{0})$ is the creep coefficient
$\sigma_{c,QP}$ is the stress in the concrete adjacent to the tendons due to prestress and the quasi-permanent actions (if appropriate)
A_{p} is the area of the prestressing tendons
A_{c} is the concrete cross-sectional area
I_{c} is the second moment of area of the concrete section
z_{cp} is the distance between the centre of gravity of the section and the tendons

Equation (4.33) is directly applicable to bonded tendons when local values of stresses are used.

4.2 Precast prestressed elements

4.2.1 Proprietary systems

Precast prestressed concrete wall panels are used in several specialist proprietary systems for the construction of circular tanks for use in clean and wastewater treatment processes, water storage and other similar applications. Tanks ranging in capacity between 50 m^3 and 12500 m^3 have been constructed successfully, both free-standing and part-buried.

Typically, these structures comprise an in-situ reinforced concrete base slab upon which the precast wall panels are erected. The panels are reinforced to resist horizontal out-of-plane bending, which may be supplemented by pre-tensioned prestressing wires in taller panels. Watertightness is achieved by circumferential post-tensioning using tendons running in horizontal ducts cast into the concrete. The vertical joint between adjacent wall panels can be made in different ways. In one proprietary system, a gap is left between each panel, which is filled with in-situ concrete. In another system, the panel edges fit close together with a rubber strip between the adjacent panel edges, which is compressed to provide a seal when the post-tensioning force is applied. The joint between the wall panels and base slab is commonly detailed as a rebate, with hydrophilic sealing strips between the base and wall panels used to make the joint watertight.

Precast concrete panel tank systems can offer advantages over steel tanks and traditional, cast in-situ concrete tanks in terms of both reduced cost and construction time. However, care is needed at all stages including design, detailing, specification, precasting, transportation, handling and erection to ensure that the desired level of structural performance and durability is achieved. For this reason this method of construction is normally only undertaken by experienced specialist companies.

4.2.2 Precast roof slabs

There is little problem in adopting such units for the roof of, say, a reservoir where structural continuity between the roof and the walls is not required. However, because of the moist humid conditions at the soffit of the roof slab, the units must be designed with no tensile stresses in the serviceability limit state. For proprietary precast units, this condition will need verifying with the manufacturer.

4.3 Cylindrical prestressed concrete tanks

4.3.1 Actions

A cylindrical tank is an efficient way of containing liquid as the radial pressure at a given depth is uniform around the tank. Thus at any level the circumferential tensile forces may be determined as a function of the depth assuming thin wall theory. The stresses induced by the prestressing will need to counteract the radial tensile forces and as a result the amount of prestress will vary with depth, being a maximum at the base of the tank (Figure 4.1). It is preferable to ensure the wall always remains in compression and the prestress levels will need over-designing as it is almost impossible to provide a totally accurate assessment of prestress losses.

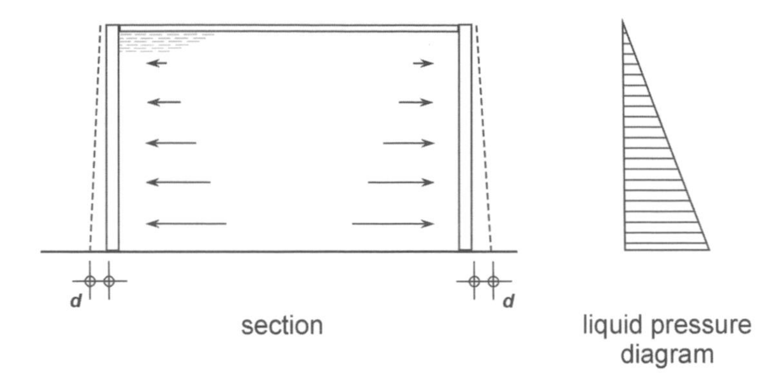

Figure 4.1 *Outward deflection of cylindrical tank (free at base).*

4.3.2 Base restraint

An additional problem arises owing to the effect of base restraint. There are three possible degrees of restraint possible at the base (Figure 4.2):

(i) Fixed: In this case there is complete moment transfer in the vertical plane between the wall and base.

(ii) Pinned: In this case the position of the base of the wall is fixed but there is no moment transfer.

(iii) Free (frictional restraint): Here the base of the wall may slide, albeit subject to a degree of frictional restraint depending upon the exact details of the joint and the vertical self weight of the wall.

Only in the last case where the wall can deform horizontally can the prestress be effectively mobilised. It is usual in such cases to ignore any effects of friction at the base.

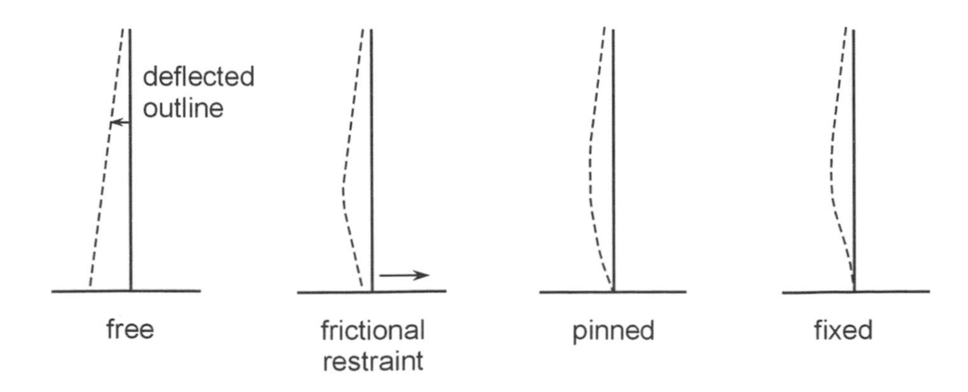

Figure 4.2 *Effect of base restraint on a loaded prestressed concrete tank.*

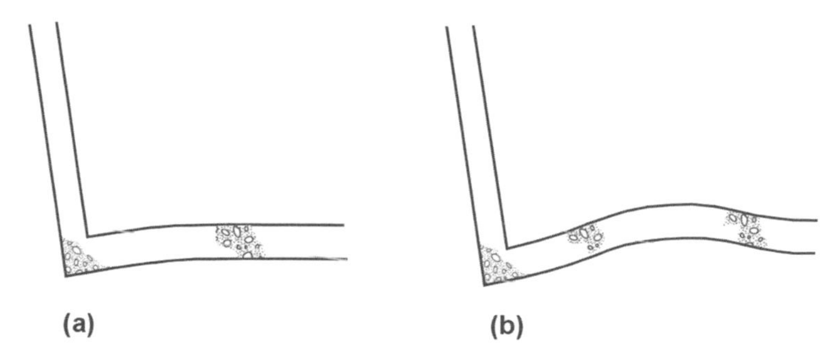

Figure 4.3 *Rotation of a 'fixed' wall footing (a) Due to settlement (b) Due to flexibility of the footing.*

Full fixity is in any case an upper bound to the real behaviour of a wall owing both to settlement from the extra loading at the edge of the base and the flexibility of the base itself (Figure 4.3). In practice it may not be possible to distinguish between the effects of settlement and base flexibility.

4.3.3 Vertical design

It is essential that the same assumptions with regard to base fixity are made for both the prestressing design in the circumferential direction and the effects of actions in the vertical direction, i.e. if the assumption of a sliding base is made in order to simplify the prestress calculations, then there can be no moments transmitted to the base due to vertical water pressure. Equally, if the base of the wall can slide there can be no moments induced in the wall due to the variation of prestress with height.

Only for deep tanks would it be economic to prestress in the vertical direction otherwise they should be designed as normal reinforced concrete. Anchor (1992) suggests that this limit occurs at around 7 m.

It is considered that the only complete solution for a circular tank prestressed circumferentially and either reinforced or prestressed in the vertical direction can be obtained using finite element techniques. Such techniques may also be extended to model the interaction between ground conditions and base fixity.

Stress analysis can provide some answers to the interaction between prestress and joint fixity.

Methods 1 and 2 below may be able to bc used to give preliminary data for design.

Method 1: Encastré base to the wall

This is of limited use as it only gives maximum values of key parameters. It does not allow either the determination of the variation of parameters with wall height nor the moments induced by prestress.

If the wall is continuous with respect to the base, then the maximum horizontal (circumferential) force does not occur at the base (Figure 4.4) but at a height of h_{max}.

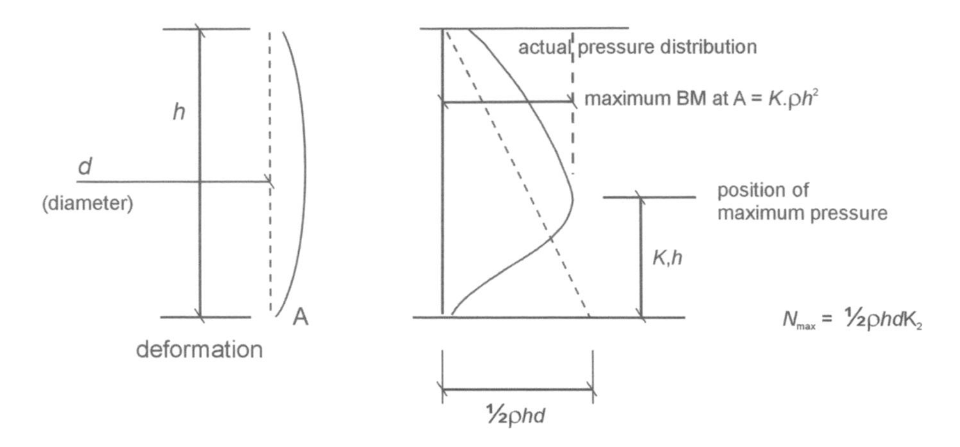

Figure 4.4 *Data for N_{max} and M_{max} (encastré base).*

The maximum circumferential tension $N_{\max}$ is given by

$$N_{\max} = \frac{1}{2}\rho K_2 dh \tag{4.34}$$

and occurs at a height $h_{\max}$ above the base given by

$$h_{\max} = K_1 h \tag{4.35}$$

The maximum bending moment at A, $M_{\max}$ is given by

$$M_{\max} = \rho K h^3 \tag{4.36}$$

where

h is the depth of the tank (or of the contained liquid)
d is the tank diameter
h_A is the wall thickness at the base of the tank
ρ is the specific weight of the contained liquid

Values of K, K_1 and K_2 are given in Figures 4.5 to 4.7 (plotted from values given in Table 154, Reynolds *et al.,* 2007). The basis of the data are not given but almost certainly are for walls of constant thickness in spite of h_A being defined as the thickness at the base of the tank.

Method 2: Analytical expressions for stress resultants due to various base fixities under hydrostatic loading

Analytical expressions for both fixed base and pinned base can be determined using a full stress analysis. The results assume the wall is of constant thickness over its height and that it may be considered as thin (i.e. any variation of stress through the wall may be ignored). The theory is subject to the value of h being large compared with $\sqrt{(0.5dh_A)}$. This condition may not be true for concrete tanks, but the equations may be used for initial estimates of internal stress resultants.

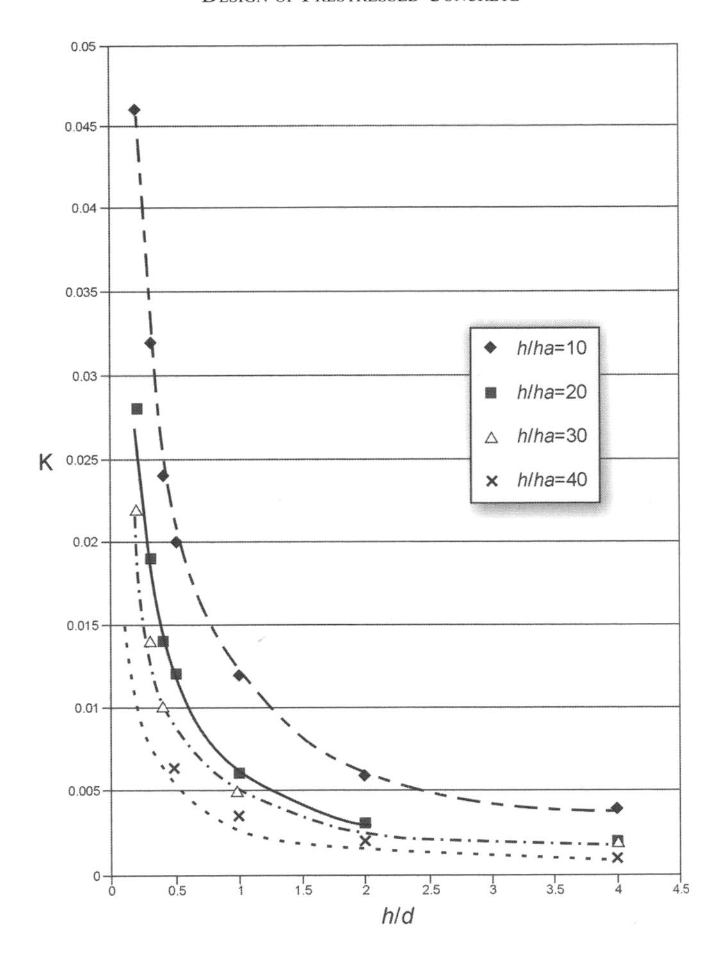

Figure 4.5 *Values of K.*

The case for the fixed base has been taken from Wang (1953) and the pinned base version has been derived from the theory therein. The notation has been amended to agree with Method 1.

In all cases a parameter βL is defined by

$$(\beta h)^4 = 3\left((1-\nu^2)\right)\left(\frac{2h}{d}\right)^2\left(\frac{h}{h_A}\right)^2 \tag{4.37}$$

where

ν is Poisson's ratio (taken as 0.2 (EN 1992-1-1 Cl 3.1.3(4)))
h is the height
d is the diameter
h_A is the wall thickness

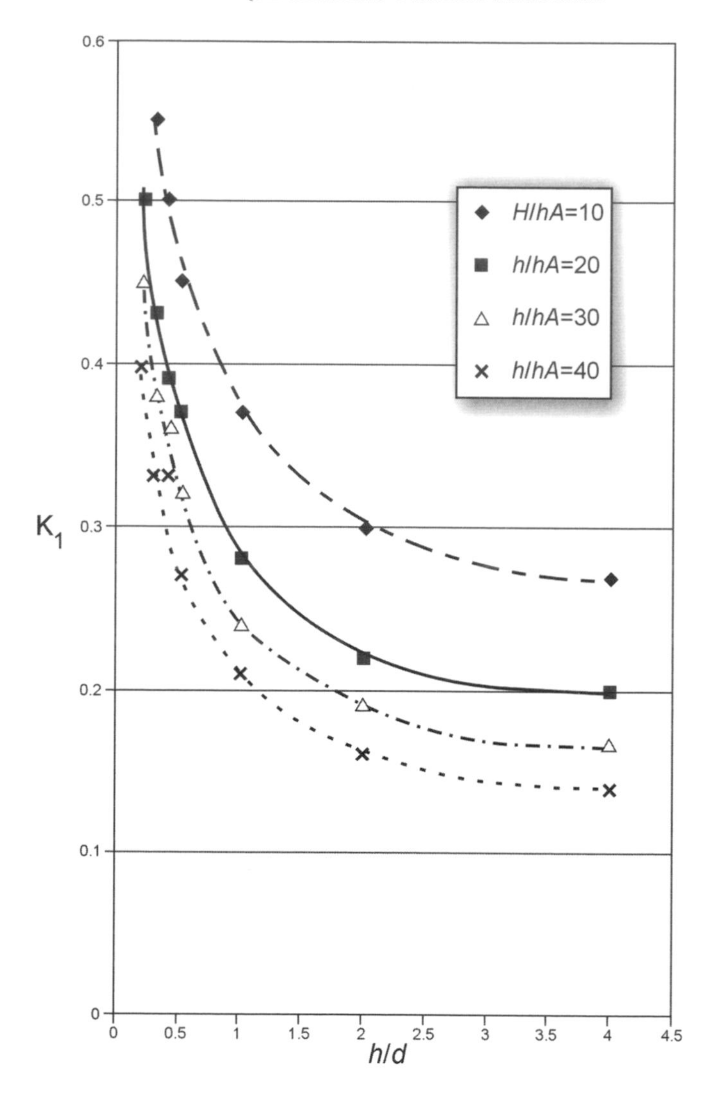

Figure 4.6 *Values of K_1.*

Fixed base

The circumferential force N_θ at a normalised height of $\frac{x}{h}$ from the base is given by

$$N_\theta = \rho ah\left\{1-\frac{x}{h}-e^{-\beta h\frac{x}{h}}\left[\cos\beta h\frac{x}{h}+\left(1-\frac{1}{\beta h}\right)\sin\beta h\frac{x}{h}\right]\right\} \qquad (4.38)$$

The vertical moment M_x at a height x from the base is given by

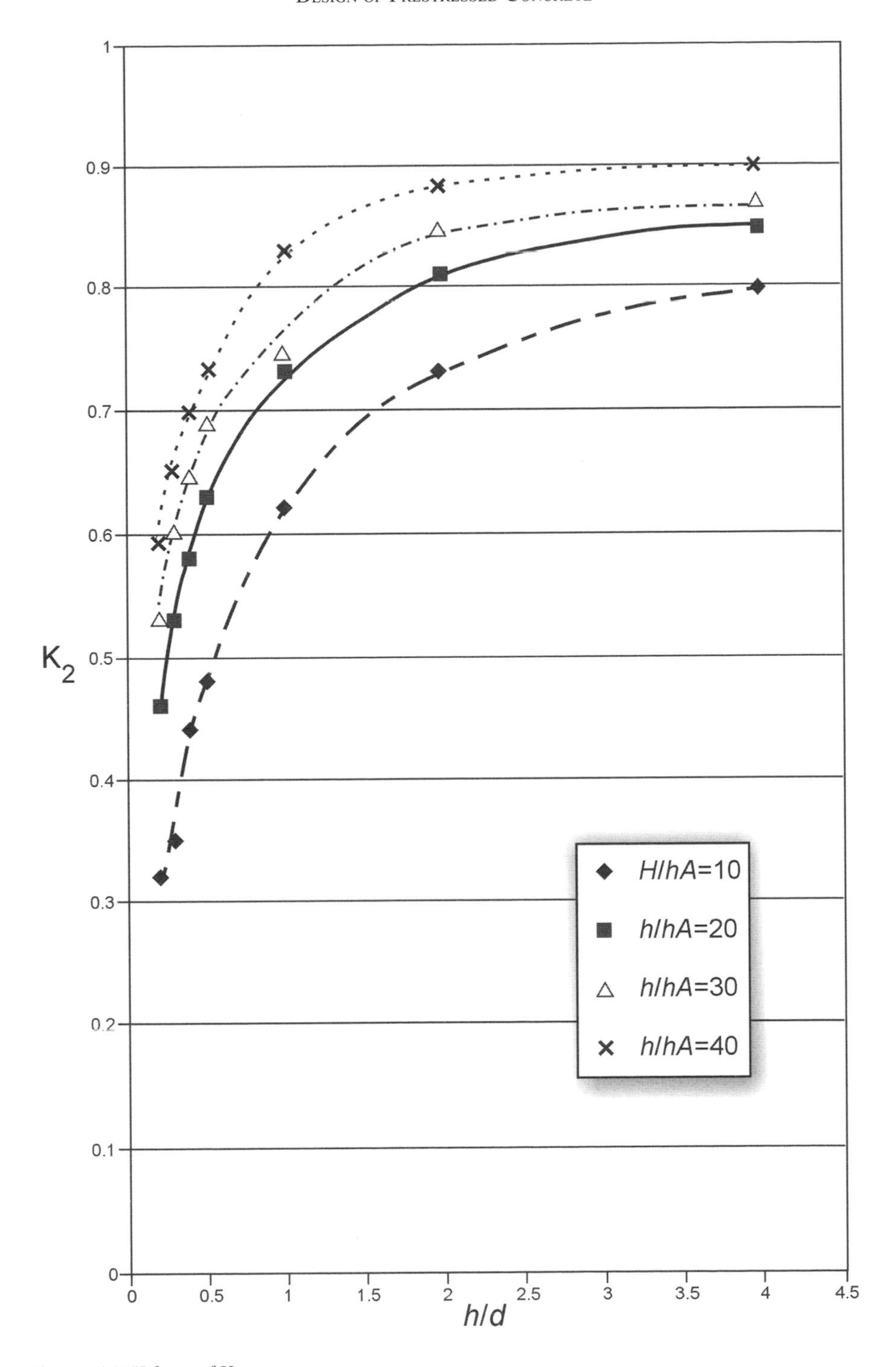

Figure 4.7 *Values of K_2.*

$$M_x = \frac{\rho \frac{d}{2} h h_A}{\sqrt{12(1-\nu^2)}} e^{-\beta h \frac{x}{h}} \left[-\sin \beta h \frac{x}{h} + \left(1 - \frac{1}{\beta h}\right) \cos \beta h \frac{x}{h} \right] \tag{4.39}$$

The circumferential moment M_θ at a height x from the base is given by

$$M_\theta = \frac{\nu \rho \frac{d}{2} h h_A}{\sqrt{12(1-\nu^2)}} e^{-\beta h \frac{x}{h}} \left[-\sin \beta h \frac{x}{h} + \left(1 - \frac{1}{\beta h}\right) \cos \beta h \frac{x}{h} \right] \tag{4.40}$$

The maximum vertical moment M_A occurs at the base of the wall and is given by

$$M_A = \frac{\rho \frac{d}{2} h h_A}{\sqrt{12(1-\nu^2)}} \left(1 - \frac{1}{\beta h}\right) \tag{4.41}$$

The maximum circumferential force N_{max} occurs at a height $\frac{x}{h}$ given by the numerical solution of

$$\beta h e^{-\beta h \frac{x}{h}} \left[\frac{1}{\beta h} \cos \beta h \frac{x}{h} + \left(2 - \frac{1}{\beta h}\right) \sin \beta h \frac{x}{h} \right] - 1 = 0 \tag{4.42}$$

Pinned base

The circumferential force N_θ at a normalised height of $\frac{x}{h}$ from the base is given by

$$N_\theta = \rho a h \left\{ 1 - \frac{x}{h} - e^{-\beta h \frac{x}{h}} \cos \beta h \frac{x}{h} \right\} \tag{4.43}$$

The vertical moment M_x at a height x from the base is given by

$$M_x = -\frac{\rho \frac{d}{2} h h_A}{\sqrt{12(1-\nu^2)}} e^{-\beta h \frac{x}{h}} \sin \beta h \frac{x}{h} \tag{4.44}$$

The circumferential moment M_θ at a height x from the base is given by

$$M_\theta = -\frac{\nu \rho \frac{d}{2} h h_A}{\sqrt{12(1-\nu^2)}} e^{-\beta h \frac{x}{h}} \sin \beta h \frac{x}{h} \tag{4.45}$$

The maximum vertical moment M_A occurs at a normalised height $\frac{x}{h}$ given by

$$\frac{x}{h} = \frac{1}{\beta h} \frac{\pi}{4} \tag{4.46}$$

and has a value of

$$M_{max} = -\frac{\rho \frac{d}{2} h h_A}{\sqrt{24(1-\nu^2)}} e^{-\frac{\pi}{4}} \tag{4.47}$$

The maximum circumferential force N_{max} occurs at a height $\frac{x}{h}$ given by the numerical solution of

$$\beta h e^{-\beta h \frac{x}{h}}\left[-\cos\beta h\frac{x}{h}+\sin\beta h\frac{x}{h}\right]+1=0 \tag{4.48}$$

Example 4.1 Design of horizontal prestress for a circular tank

Design the external prestressing for a circular tank 20 m diameter with a height of water 7.5 m with a 0.5 m freeboard assuming the wall is free to slide on the base. The thickness of the wall is 250 mm (Figure 4.8).

Materials:

Concrete grade: C35/45

Prestressing strand: 15.2 mm diameter low relaxation. Characteristic 0.1% proof load is 197 kN.

Assume the prestressing operation is carried out at 7 days after casting and the initial prestress is taken as 80% of the breaking load.

$$f_{pi}=\frac{0.8\times197\times10^3}{\pi\frac{15.2^2}{4}}=869\ MPa$$

From typical manufacturers' data the 1 000 hr relaxation is 4.5%.

The amount of prestress has to be estimated before a complete set of calculations can be carried out. Assume 40% loss in prestress at serviceability and design for the stress due to the water pressure and an additional 1 MPa to ensure the section is under compression. Estimate of prestressing force P_{eff}:

$$P_{eff}=0.8\times(1-0.4)\times197=95\ kN$$

The force to be resisted is calculated using 7 No 1 m high strips and a residual strip of 0.5 m at the top. The total stress resultant to be resisted is calculated at the bottom of each strip. To ensure the structure remains uncracked, the stress resultant

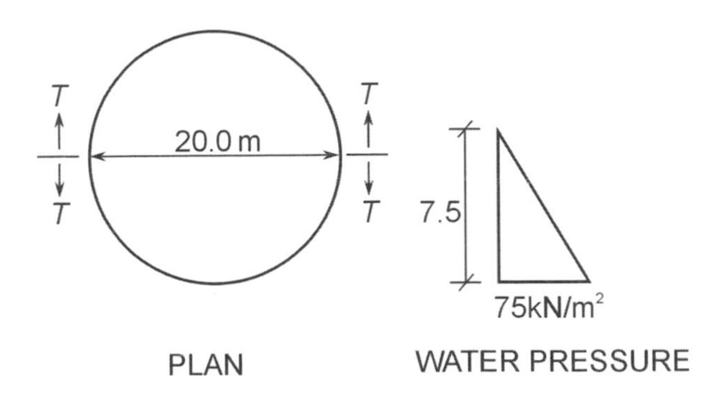

Figure 4.8 *Design data.*

is taken as that due to water pressure and additionally due to a stress of 1 MPa in the concrete.

The stress resultant N_p due to the water pressure is given by

$$N_p = \rho(7.5 - x)\frac{D}{2} \tag{4.49}$$

where

ρ is the specific weight of water (10 kN/m^3)
x is distance from the base of the tank to the bottom of the strip being considered
D is the diameter of the tank

The stress resultant N_σ due to the unit stress of 1 MPa is given by

$$N_\sigma = \sigma w t \tag{4.50}$$

where σ is the value of the applied stress (1 MPa), w is the width of a strip and t is the wall thickness.

The stress resultants N_p and N_σ are given in Table 4.3 and Figure 4.9.

Stressing sequence

The tendons are anchored on concrete pilasters at the quarter points around the tank with the cables extending half-way round the tank (Figure 4.10). Thus the first cable is anchored at B and stressed at A and C. Simultaneously, another cable anchored at D is also stressed at A and C (hence four jacks are required). The next pair of cables up would be stressed at B and D and anchored at C and A. It is important that the stressing is performed sensibly and uniformly up the tank otherwise there will be an additional stress not considered in the design. This will happen if say one half of the tank (vertically) is stressed before the other half.

Table 4.3 *Values of applied stress resultants and estimates of prestress.*

Strip No	*height*	*Force due to water*	*Additional force*	*Total*	*No of tendons*	*Spacing*
	m	*kN*	*kN*	*kN*		*mm*
1	0–1	750	250	1000	11	90
2	1–2	650	250	900	9	110
3	2–3	550	250	800	8	125
4	3–4	450	250	700	7	140
5	4–5	350	250	600	6	165
6	5–6	250	250	500	5	200
7	6–7	150	250	400	4	250
8	7–7.5	50	125	175	2	250

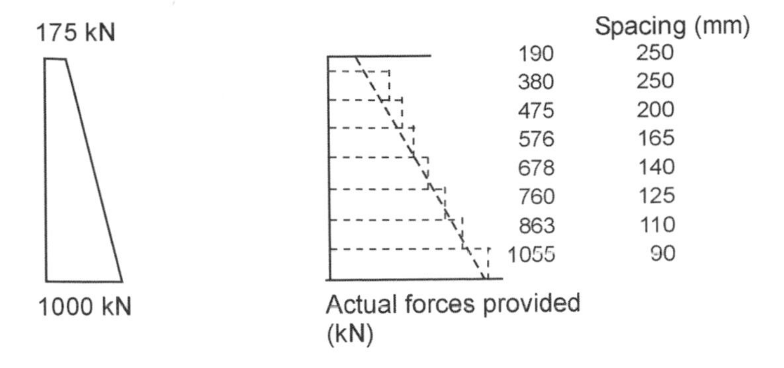

Figure 4.9 *Applied actions and tendon forces.*

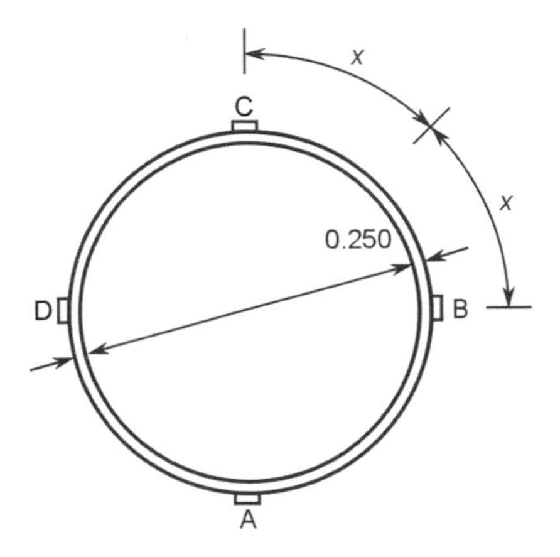

Figure 4.10 *Stressing details.*

Friction losses

The circumference of the tank L is given by

$$L = \pi(D + 2t) = \pi(20 + 2 \times 0.25) = 64.4\,m$$

The cables are half this length and are stressed from both ends (and are anchored in the centre); then the maximum friction losses occur at $L/8$ or 8.05 m.

Radius of curvature of the tendon is 10.5 m.

Take k as 0.0075 and μ as 0.24.

The angle θ is given by

$$\theta = \frac{x}{R} = \frac{8.05}{10.5} = 0.767 \quad radians$$

Determine percentage loss from Eq. (4.32)

$$\Delta f_s = 100\left(1 - e^{-\mu(\theta + kx)}\right) = 100\left(1 - e^{-0,24(0,767+0,0075\times 8,05)}\right) = 18\%$$

Elastic losses

At transfer (and at serviceability):

The concrete strength $f_{cm}(t)$ (Cl 3.1.2(6) BS EN 1991-1-1) is given by

$$f_{cm}(t) = f_{cm}\exp\left\{s\left[1 - \sqrt{\frac{28}{t}}\right]\right\}$$

From Table 3.1 (BS EN 1992-1-1), f_{cm} for C35/45 is 43 MPa (i.e. $f_{ck} + 8$).

Assume cement class CEM 32.5 (Class N); thus from Table 4.1, s = 0.25, so $f_{cm}(7)$ is given as

$$f_{cm}(7) = 43\exp\left\{0.25\left[1 - \sqrt{\frac{28}{7}}\right]\right\} = 33.5\ MPa$$

Determine the elastic modulus at 7 days, $E_{cm}(7)$

$$E_{cm}(7) = 22\left(\frac{f_{cm}(7)}{10}\right)^{0.3} = 22\left(\frac{33.5}{10}\right)^{0.3} = 31.6\ kPa$$

The stress in the concrete adjacent to the prestressing strand, σ_{cp}, is determined by considering the axial effect of the prestress. There is no moment induced by the eccentricity of the prestress with respect to the centroid of the section as the structure must deform symmetrically.

$$\sigma_{cp} = \frac{P}{A_c}$$

$$A_c = 250 \times 1000 = 0.25 \times 10^6\ mm^2$$

$$P = 11 \times (0.8 \times 197) = 1733\ kN$$

$$\sigma_{cp} = \frac{P}{A_c} = \frac{1733 \times 10^3}{0.25 \times 10^6} = 6.93\ MPa$$

From Eq. (4.4) the percentage elastic loss is given by

$$\Delta f_s = 100\frac{\sigma_{cp}}{f_{pi}}\frac{E_s}{E_{cm}} = 100\frac{6.93}{869}\frac{195}{31.6} = 4.9\%$$

As the tank is post-tensioned, this value may be halved to give 2.5%.

Shrinkage losses:
Autogenous shrinkage (Eq. (4.6)):

$$\varepsilon_{ca}(t) = \left[2.5(f_{ck} - 10)\times 10^{-6}\right]\left[1 - \exp\left(-0.2t^{0.5}\right)\right]$$

At serviceability, t tends to infinity, thus

$$\varepsilon_{ca}(\infty) = 2.5(f_{ck} - 10)\times 10^{-6}$$

Thus

$$\varepsilon_{ca}(\infty) = 2.5(f_{ck} - 10)\times 10^{-6} = 2.5(35 - 10)\times 10^{-6} = 63\times 10^{-6}$$

Drying shrinkage (Eq. (4.7))
From Eq. (4.15) determine h_0

$$h_0 = \frac{2A_c}{u} = \frac{2\times 7500\times 250}{250 + 2\times 7500} = 246 \ mm$$

Determine k_h using Eq. (4.12) as $200 < h_0 < 300$ mm,

$$k_h = 0.85 - 0,001(h_0 - 200) = 0.85 - 0.001(246 - 200) = 0.804$$

Determine β_{RH} from Eq. (4.9) with $RH = 85\%$,

$$\beta_{RH} = 1.55\left[1 - \left(\frac{RH}{RH_0}\right)^3\right] = 1.55\left[1 - \left(\frac{85}{100}\right)^3\right] = 0.598$$

Determine $\varepsilon_{cd,0}$ from Eq. (4.8) with Class N cement ($\alpha_{ds1} = 4$ and $\alpha_{ds2} = 0.12$, both from Table 4.1)

$$\varepsilon_{cd,0} = 0.85\left[(220 + 110\alpha_{ds1})\exp\left(-\alpha_{ds2}\frac{f_{cm}}{f_{cm0}}\right)\right]\beta_{RH}\times 10^{-6}$$

$$= 0.85\left[(220 + 110\times 4)e^{-0,12\frac{35}{10}}\right]\times 0.598\times 10^{-6} = 220.4\times 10^{-6}$$

At serviceability take $\beta_{ds}(t,t_s) = 1.0$.
From Eq. (4.7), $\varepsilon_{cd}(\infty)$ is given by

$$\varepsilon_{cd}(\infty) = \beta_{ds}(t,t_s)k_h\varepsilon_{cd,0} = 1.0\times 0.804\times 220\times 10^{-6} = 177\times 10^{-6}$$

The total shrinkage ε_{cs} is given by Eq. (4.16) as

$$\varepsilon_{cs} = \varepsilon_{cd} + \varepsilon_{ca} = 177\times10^{-6} + 63\times10^{-6} = 240\times10^{-6}$$

The loss in prestress is given by Eq. (4.17)

$$\Delta f_s = 100\frac{E_s\varepsilon_{cd,o}}{f_{pi}} = 100\frac{195\times10^3\times240\times10^{-6}}{869} = 5.4\%$$

Creep:
At serviceability, $\beta_c(\infty,t_0) = 1.0$.
Determine $\beta(f_{cm})$ from Eq. (4.24),

$$\beta(f_{cm}) = \frac{16.8}{\sqrt{f_{cm}}} = \frac{16.8}{\sqrt{35}} = 2.84$$

Determine $\beta(t_0)$ from Eq. (4.25) with $t_0 = 7$ (days)

$$\beta(t_0) = \frac{1}{0.1+t_0^{0.20}} = \frac{1}{0.1+7^{0.2}} = 0.635$$

As $f_{cm} = 35$ MPa, use Eq. (4.22) to determine φ_{RH} ($h_0 = 246$ mm from above) with $RH = 85\%$,

$$\varphi_{RH} = 1+\frac{1-0.01(RH)}{0.1\sqrt[3]{h_0}} = 1+\frac{1-0.01\times85}{0.1\sqrt[3]{246}} = 1.24$$

Determine φ_0 from Eq. (4.21),

$$\varphi_0 = \varphi_{RH}\beta(f_{cm})\beta(t_0) = 1.24\times2.84\times0.635 = 2.24$$

Determine $\varphi(\infty,t_0)$ from Eq. (4.20),

$$\varphi(\infty,t_0) = \varphi_0\beta(\infty,t_0) = 2.24\times1.0 = 2.24$$

Assuming no slip between the tendons and the concrete, from strain compatibility the strain in the tendons is also equal to ε_{cs} and the percentage loss in prestress Δf_s is given by Eq. (4.19) as

$$\Delta f_s = 100\varphi(\infty,t_0)\frac{\sigma_c}{f_{pi}}\frac{E_s}{E_{cm}} = 100\times2.24\frac{6.93}{869}\frac{195}{31.6} = 11.0\%$$

Anchorage loss:
The loss due to bedding in at the anchorages will be taken up by increasing the prestress force.

Total losses (in per cent):

Relaxation	4.5
Friction	18.0
Elastic	2.5
Shrinkage	5.4
Creep	11.0
Total	**41.4**

This is sufficicntly closc to thc assumcd valuc of 40% to not nccd a rcdcsign of prestress.

Check the losses using the formula in BS EN 1992-1-1.

Rewrite the formula in Eq. (4.33) to give the direct loss in stress as

$$\sigma_{p,c+s+f} = \frac{\varepsilon_{cs} E_p + 0.8\Delta\sigma_{pr} + \frac{E_p}{E_{cm}}\varphi(t,t_0)\sigma_{c,Qp}}{1+\frac{E_p}{E_{cm}}\frac{A_p}{A_c}\left(1+\frac{A_c}{I_c}z_{cp}^2\right)\left[1+0.8\varphi(t,t_0)\right]}$$

$$\varepsilon_{cs} = 240\times10^{-6}$$

$$\varphi(t,t_0) = \varphi(\infty,t_0) = 2.24$$

$$z_{cp} = 0$$

$$A_c = 0.25\times10^6 \quad mm^2$$

$$A_p = 11\times\pi\left(\frac{15.2}{2}\right)^2 = 1996 \ mm^2$$

$$E_{cm} = 31.6 \ kPa$$

$$\Delta\sigma_{pr} = 0.045\times f_{pi} = 0.045\times869 = 39 \ MPa$$

$$\sigma_{p,c+s+f} = \frac{\varepsilon_{cs} E_p + 0.8\Delta\sigma_{pr} + \frac{E_p}{E_{cm}}\varphi(t,t_0)\sigma_{c,Qp}}{1+\frac{E_p}{E_{cm}}\frac{A_p}{A_c}\left(1+\frac{A_c}{I_c}z_{cp}^2\right)\left[1+0.8\varphi(t,t_0)\right]}$$

$$= \frac{240\times10^{-6}\times195\times10^{3}+0.8\times39+\dfrac{139}{31.6}\times2.24\times6.93}{1+\dfrac{195}{31.6}\dfrac{1996}{0.25\times10^{6}}(1+0.8\times2.24)} = 128.6\ MPa$$

$$\text{Frictional loss} = \frac{128.6}{869} = 0.148 \text{ or } 14.8\%$$

The losses due to friction must be added to this; thus the total loss is $18.0 + 14.8 = 32.8\%$. This is lower than that calculated from first principles.

Chapter 5
Distribution reinforcement and joints: Design for thermal stresses and shrinkage in restrained panels

There are many situations within normal concreting practice when cracking due to the restraint of imposed deformations (i.e. early thermal movement and shrinkage) may be difficult to avoid. In fact, cracking from the restraint of early thermal movements (often referred to as 'non-structural' cracking) is the most common form of restraint induced cracking (Bamforth, 2007; Harrison, 1991). The fact that cracks can occur is not a problem provided there is a robust design approach to controlling these cracks. The approach for the design of reinforcement to control early age thermal cracking presented in the codes for the design of water-retaining structures, which has been used for over 30 years, has generally been successful, and this approach has broadly been adopted in BS EN 1992-3. However, field observations have identified crack widths in excess of those predicted by the code and it is apparent that these cases of 'non-compliance' are not attributable to anomalous behaviour but are a result of the basic assumptions behind the current design approach being incorrect. In particular:

- The boundary (restraint) conditions play a more significant role in determining the crack pattern and widths–the reality is that walls are more likely to be restrained along their base rather than at their ends, which is the current restraint condition on which the code guidance is based.
- For edge-restrained conditions the full crack pattern in BS EN 1992-3 rarely occurs.

The question is why has the approach worked for the majority of cases when the basic assumptions on which it is based appear to be incorrect? The answer probably is down to more luck than judgement. Significant work has been performed at the École Polytechnique Fédérale de Lausanne, which shows that for end restraint and for the levels of imposed deformation likely in practice (strains less than 1000×10^{-6}) the current analysis (based on end restraint) for crack width is in fact incorrect as the width depends not on the imposed strain but upon the tensile strain capacity of the concrete. It should also be noted that the spacing of the cracks predicted by the current analysis bears no relationship to the spacings obtained by the extensive tests carried out by Farra and Jaccoud (1993). The fact is that in all likelihood possibly double the amount of steel is required by the current end only restraint based guidance, which possibly explains the success of the current approach in a lot of the cases, i.e. too much steel.

As mentioned in Chapter 1, since the edge-restrained situation is probably the most common in practice, or at least practice will tend to suggest a combination of the two restraint conditions, it is clear that to present design guidance based only on the analysis of end restraint is probably wrong and logically indefensible. In the approaches of both BS 8007 and BS EN 1992-3, the restraint is assumed to influence only the magnitude of restrained strain but not the crack spacing; hence, higher restraint leads to wider cracks. However, limited evidence suggests that the restraint also influences the crack spacing and the extent to which cracks may open; a proposal by Bamforth *et al.*, (2010) suggests that higher restraint may lead to higher cumulative crack width, but distributed as finer cracks. There are also unpublished analyses by Beeby that appear among Eurocode committee documents (2000–2005) and work by Kheder (1997) and Al Rawi and Kheder (1990). All of this work suggests that the amount of reinforcement provided has a relatively smaller influence on the crack widths than is the case in other forms of cracking. The concept of minimum reinforcement ratio is also largely irrelevant as the stress in the steel appears to be relatively low.

Whilst end restraint has been investigated a great deal, edge (base) restraint has been studied very little. In fact only two experimental investigations can be cited where base restraint has been researched (Nilsson, 2000 & 2003), and unfortunately the experimental materials, approach and scarcity of reported information mean it is difficult to conclude much from these investigations. However, there is sufficient academic thinking and field observations to support this range of different restraint types and combinations (i.e. edge or end or a combination of both) such that the idea of edge restraint was incorporated in Annex M of BS EN 1992-3. Its inclusion highlights the issue to the designer and the approach has been developed further by Bamforth in the updated version of the CIRIA document, *Early-age Thermal Crack Control in Concrete* (2007).

5.1 Cracking due to different forms of restraint in reinforced concrete

Non-structural cracking results from either internal restraint to differential expansion / contraction or external restraint to contraction.

5.1.1 Internal restraint

If a small cube of concrete is cast (such as a test cube), it is unlikely to exhibit surface cracking or any internal cracking as the differential expansion / contraction between the surface and the core of the cube will be small. However, in larger dimension concrete elements (typically structural), differential temperature changes can occur (the stimulus for these changes in temperature is the release of the heat of hydration from the cementitious binder when the concrete is cast), which can lead to both surface and internal cracks forming (see section below). This type of cracking is known as early thermal cracking.

If a long specimen with relatively small cross-sectional dimensions (similar to the small cube above) is made, containing reinforcement, and the ends of the reinforcement are held to prevent any movement, after a few days it will be found that fine lateral cracks are present (Figure 5.1). The spacing and size of these cracks (i.e. whether there are a few wide cracks or a larger number of fine cracks) depends on the relation between the quantity of reinforcement, the bar size, the cross-sectional area of concrete and the cover. The cracks are induced as the reinforcement provides an internal restraint to the strains in the concrete caused by chemical hydration of the cement in

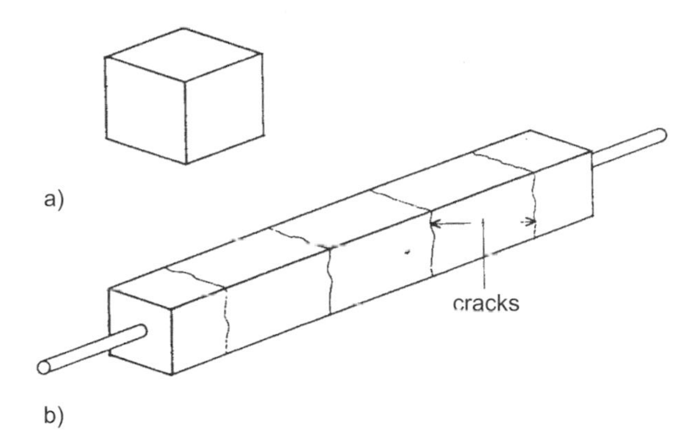

Figure 5.1 *Cracking in concrete elements (a) Small cube free to move (b) Long reinforced element restrained by reinforcement.*

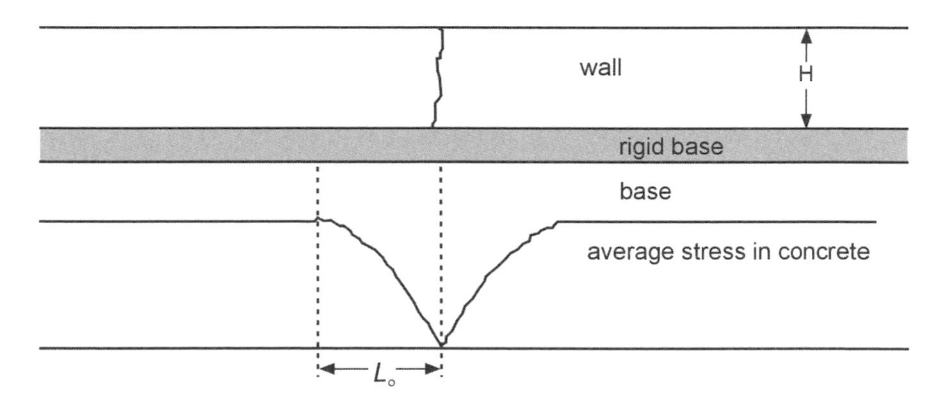

Figure 5.2 *Conditions in a base-restrained wall after initial cracking.*

the concrete mix in the short term and potentially long term by restraining the time-dependent movements of the concrete.

5.1.2 External restraint

As mentioned previously, external restraint to contraction can exist in two forms, base (edge) or end restraint. The factors that influence cracking in base-restrained walls are different than those in end-restrained sections. Initially, consider the way a base-restrained wall will crack and the size of the crack widths. Figure 5.2 illustrates the conditions in a wall after initial cracking.

Initially, consider a wall completely restrained along its base and with no reinforcement. Once sufficient thermal movement has occurred, cracks will form. At the primary crack (which will be the full height of the wall) the stress in the concrete is zero. However, with increasing distance from the crack, stress is transferred to the wall by shear at the interface with the base until, at some distance L_0 from the crack, the stress distribution is unaffected by the crack. In the end-restrained case, the crack reduces the stiffness of the whole system and hence reduces the stresses throughout (the effect of the crack is global). This is fundamentally different to the edge-restrained case considered here where the relief of stress caused by the primary crack is purely local (secondary cracks may form depending on the level of stress relief provided by the primary cracks–see section 5.3). Stresses are not relieved beyond L_0 from the primary crack and further primary cracks may form in the unrelieved areas. These other primary cracks will have no influence on the first crack.

5.2 Causes of cracking

There are several forms of movement that can occur in concrete. Some are short-term effects such as the expansion / contraction from the heat generated during hydration of the cement paste and the contraction from autogenous shrinkage. Drying shrinkage (as well as autogenous shrinkage to a lesser effect) and environmental effects are long-term effects. To be clear, any movement that is unrestrained will not generate cracks. Cracks are formed because these movements are restrained.

5.2.1 Short-term movements

Heat of hydration

When materials are mixed together to make concrete, a chemical reaction takes place between the cement and water during which heat is evolved. This heat of hydration causes the temperature of the concrete to rise. A typical curve illustrating the temperature rise and fall in concrete during the first few days after mixing is shown in Figure 5.3(a). The peak temperature depends on several factors (i.e. insulation, environment, cement content, etc.) and is normally reached between 1 and 3 days (steps can be taken to reduce this peak). After this time, about 50% of the total heat will have been liberated. The peak temperature is reached when the rate of heat generation reduces to such an extent that heat loss exceeds heat generation. At this point the concrete cools, as heat is dissipated to the environment, and the concrete contracts. Typically, heat is still being lost after 6 months (at 6 months approximately 90% of the total heat generated has been lost).

According to traditional theory, during the period when the concrete temperature is increasing, expansion will take place. If the expansion is restrained by adjoining sections of hardened concrete, some creep will occur in the relatively weak concrete, relieving the compressive stresses induced by the attempted expansion. As the concrete subsequently cools, it tries to shorten but, if there are restraints present, tensile strains will develop leading to cracking (Figure 5.3(b)). This is known as 'early thermal movement' and assumes that external restraint dominates. The fundamental theory of early thermal and long-term cracking of sections from 300 mm to over 1000 mm due to imposed deformations is discussed further in section 5.3.4. (It is still good practice that formwork should not be removed for 3 or 4 days, otherwise cold winds may cause surface cracking of the warm concrete.)

Autogenous shrinkage

Whereas drying shrinkage is a result of the loss of moisture from the concrete to the surrounding environment (see later), autogenous shrinkage is a result of the internal consumption of water from the capillary pores, which occurs during the hydration of the cement. These products of hydration occupy less volume than the sum of the

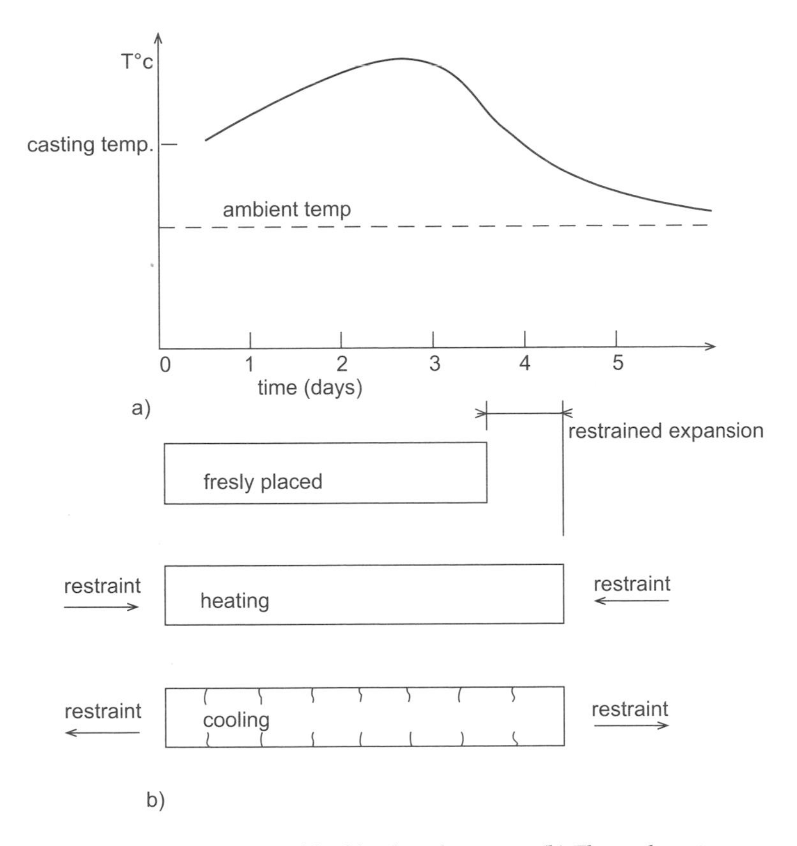

Figure 5.3 *(a) Rise in temperature of freshly placed concrete (b) Thermal strains.*

original water and unhydrated cement. The process is also known as 'self-desiccation'. As autogenous shrinkage occurs after setting, it is restrained by the aggregate particles and the hydrated cement paste 'skeleton'. It is therefore reasonably small (less than approximately 100 microstrain) in normal strength concrete (< C60/75 MPa). However, in high strength concrete or with any concrete with a low water / binder ratio it can be equivalent to or even exceed drying shrinkage.

Prior to the introduction of BS EN 1992-1-1, methods for estimating autogenous shrinkage had not been provided. BS EN 1992-1-1 not only provides a method for estimating autogenous shrinkage (based on the strength class of the concrete) but it also provides a time function so that the rate at which this shrinkage develops can be determined. The introduction of this guidance in BS EN 1992-1-1 is due to the belief that, although the quantities are small, the level of this type of shrinkage can still enhance the early age internal tensile stresses that are produced on cooling, post peak temperatures.

The total long-term shrinkage is a combination of drying and autogenous shrinkage. Drying shrinkage is not incorporated in early age calculations as this type of strain develops slowly (it is a function of the migration of moisture to the surrounding environment). To reiterate, autogenous shrinkage, on the other hand, is a strain that develops predominantly during hardening of the concrete, which occurs mostly during the early days after casting for normal strength concrete.

5.2.2 Long-term movements

Drying shrinkage

Guidance on the development of drying shrinkage with time (based on the ambient relative humidity) is also provided in BS EN 1992-1-1 for strengths up to C90/105 (CEM 1 Class N). The total long-term shrinkage is actually a combination of drying and autogenous shrinkage. However, whilst it is possible to measure autogenous shrinkage using sealed samples (i.e. samples that will not allow an exchange of moisture between the concrete and the surrounding environment) it is not possible to quantify autogenous shrinkage when samples are allowed to dry, i.e. measurement of drying shrinkage intrinsically accounts for autogenous shrinkage. In contradiction to this theory, however, BS EN 1992-1-1 proposes that total shrinkage is composed of drying shrinkage and autogenous shrinkage, i.e. that these two types of shrinkage are additive. Assessment of crack width and spacing presented later in this chapter and in Chapter 6 follows this code guidance and that presented by Bamforth (2007). The design output produced based on this guidance is therefore clearly potentially conservative.

As concrete hardens and dries out, it shrinks. This is not an irreversible process. If the concrete is subsequently placed in water (or subjected to a higher humidity) the hardened cement paste will absorb water. This reversible moisture movement can represent 40 to 70% of the drying shrinkage that has occurred prior to this increase in humidity. Practically, this is important, especially when considering the age of the concrete before water testing and subsequent commissioning. Also, the designer may need to consider any shrinkage differentials that could develop across the width of certain structural sections such as walls and roofs. For these sections, the inner face may be subjected to 100% humidity (i.e. under water), whereas the outer face may be subjected to lower humidities, depending on whether it is sealed or not and to what extent it is sealed.

If a reinforced concrete member is considered under no external stress (Figure 5.4), as mentioned previously, it will be apparent that any free shrinkage of the concrete will be prevented by the steel reinforcement. The steel is therefore put into compression and the concrete into tension, with longitudinal bond forces present on the surface of the reinforcement. The magnitude of these forces (compression, tension and bond) is dependent on the concrete properties and the ratio of the area of steel to the area of concrete.

If a high ratio of steel is present and there is no external restraint applied to the element, no cracks will form. However, if the steel ratio is relatively small or external

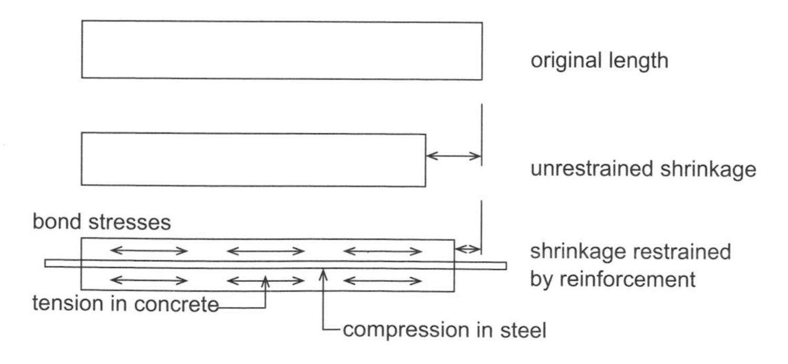

Figure 5.4 *Drying shrinkage in reinforced concrete.*

restraints are present, cracking is certain. The cracks may form at close centres and be fine in width, or may be further apart and be correspondingly wider in order to accommodate the total strain (see Figure 5.5). An indication of the crack patterns in edge-restrained walls (spacing, height, width and relationship to length (specifically height to length ratio of the panel)) is provided in Section 5.3.

Environmental conditions

An elevated concrete water tower will be subjected to strains due to changes between summer and winter temperatures. In terms of crack spacing and crack width calculations this change in ambient conditions is accounted for using the term T_2 (annual temperature change–see Section 5.3.3). However, in temperate climates (such as the UK), it is not usual to consider such temperature effects on the structural performance of normal types of water-retaining structures.

In countries where temperatures are more extreme, some allowance may need to be made (Forth *et al.*, 2005). In these countries it will also be the case that temperature differentials across the structure can be produced by the direct action of the sun (solar radiation) leading to differential strains, which may enhance stresses calculated from the normal gravity loading system (self weight, water, soil).

In the UK, solar radiation is attenuated in buried or partially buried rectangular service reservoirs using typically a 300 mm thick gravel layer on the roof of the reservoir and soil battering the sides of the structure. This is to limit temperature differentials within the structure, specifically between the roof and the wall elements. It is also to maintain the temperature within the stored area to below a certain temperature to negate any risk of bacteriological breakout within the tank.

Again, in rectangular service reservoirs within the UK, typically a sliding roof joint detail (see Section 5.4.2) is adopted (no design guidance is available with

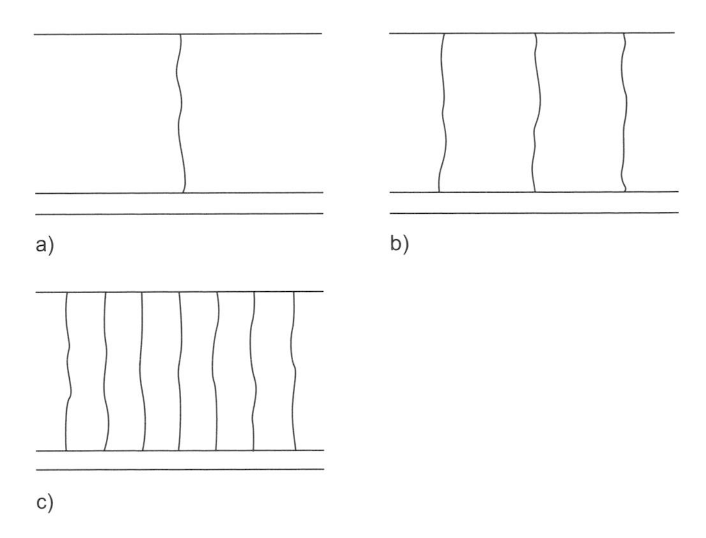

Figure 5.5 *Development of cracks related to reinforcement in a wall (a) Insufficient steel – wide cracks, steel yields (b) Controlled cracks – average spacing and width (c) High proportion of steel – well-controlled cracks of narrow width and close spacing.*

respect to the use of monolithic joints at this location). This ensures that no effect of temperature differentials is carried into the wall from the roof and that they need not be considered in design. However, examples of the effect of solar radiation can clearly be seen in some established reservoirs where the roof has 'sailed' over the wall, projecting several millimetres.

Possibly, for the last 10 to 15 years, several attempts have been made to introduce a fully monolithic roof / wall joint or at least a partially monolithic joint. This has the benefit of negating maintenance and eliminating any point of ingress for contaminants. There is minimum design guidance available for this type of joint in this type of structure and progress has been made mostly on the basis of field observations. However, research work carried out by the authors (Forth *et al.*, 2005; Muizzu, 2009; Forth, 2012) has indicated that if this monolithic joint is to be introduced, it is imperative that temperature differentials are considered and that thermal modelling is performed; typically it was shown that the joint moment was enhanced due to creep and swelling but that these additional moments reached a peak after approximately 4 years. Practically, (and from an economic perspective) it would be realistic to include additional steel to control these enhanced moments

5.3 Crack distribution

It is accepted that cracking in elements subjected purely to end restraint, due to short-term movements, may be controlled by reinforcement (Figure 5.6). The objective is to distribute the overall strain in the element between reinforcement and movement joints, so that the crack widths are acceptable or, if considered desirable, that the concrete remains uncracked. There is no single solution to the design problem of controlling short-term movements, i.e. the designer may choose to have closely spaced movement joints with a low ratio of reinforcement, or widely spaced joints with a high ratio of reinforcement. The decision is dependent on the size of the structure, method of construction to be adopted, and economics.

However, crack formation in a purely edge-restrained concrete wall is expected to depend more on the amount of restrained strain. The restrained strain is simply the amount of potential free strain (unrestrained movement) less the amount of strain offered by the restraint condition. It is suggested that the widest crack will form at

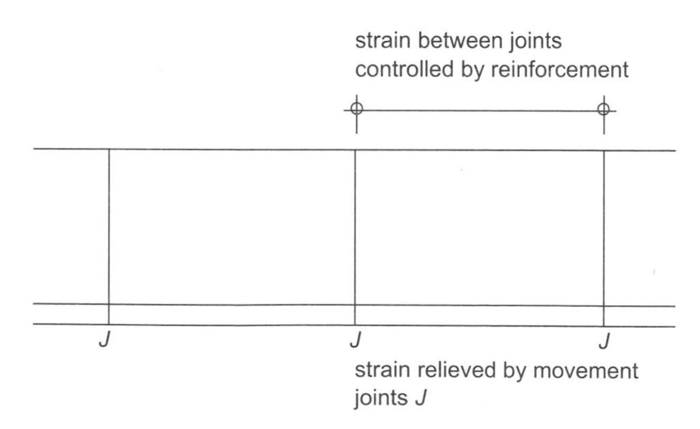

Figure 5.6 *Relation of movement joints and reinforcement in controlling strain in a wall.*

the middle of the wall, with narrower cracks forming within zones of lower restraint. Where the restraint is sufficiently low, no cracks are expected to form. Importantly, it is suggested that the amount of reinforcement provided has a relatively smaller influence on crack width than is the case in other forms of cracking (i.e. end-restrained condition). The concept of minimum reinforcement ratio is, therefore, largely irrelevant as the stress in the steel appears to be relatively low. The limited evidence available also suggests that the type of restraint also influences crack spacing. Bamforth *et al*., (2010) even proposes that higher restraint may lead to higher cumulative crack width, but distributed as more finer cracks.

Figure 5.7 is an extract from an investigation into base-restrained walls by Kheder (1997). This particular wall exhibited two primary cracks and then a number of secondary cracks. The extent of these secondary cracks was influenced by the stress relief provided by the primary crack and zones of influence were suggested, emanating from the position of the primary crack, which dictated the height and width of these secondary cracks. For the case of pure edge restraint, the stress relief offered by the crack was greatest at the top of the wall. Kheder looked at a series of walls with different length to height ratios (*L/H* ratios) and steel ratios and made a series of observations:

- For the same percentage steel, S_{max} increased with increasing *L/H*.
- As *L/H* increased, the height of the cracks increased, irrespective of steel percentage, up to an *L/H* of 4.

In summary, the combined effects of the base and the steel reinforcement determine the crack spacing in edge-restrained walls. As the wall height increases, the effect of the base decreases, i.e. for walls with the same *L/H* ratio, more widely spaced cracks and wider cracks will be seen in higher walls and, therefore, to control these cracks higher steel ratios will be required.

As mentioned in Chapter 1, it is reasonable to expect that for a wall constructed, say, between two existing panels, the restraint is due to a combination of both edge and end restraint with each type dominating at different locations, i.e. edge restraint

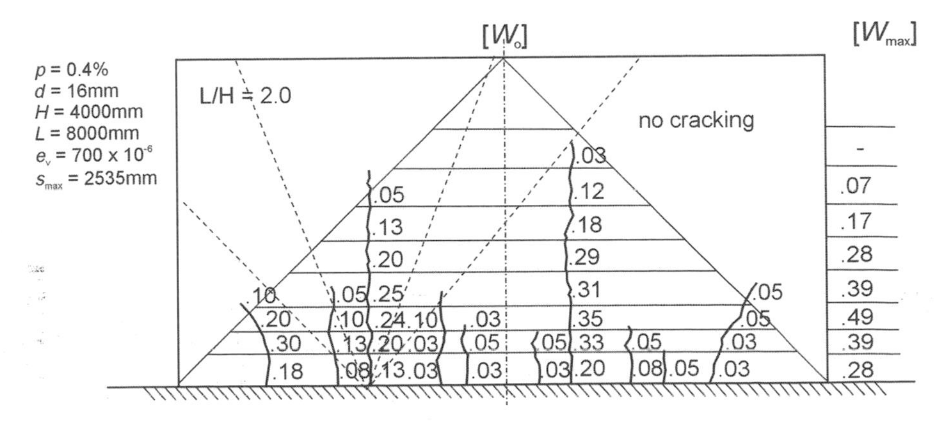

Figure 5.7 *Comparison between observed (W_o) and calculated (W_{max}) crack widths in a wall.*

dominating at the base and end restraint dominating nearer the top and towards the sides of the newly cast wall, and at different times in the life of the structure. This is, therefore, a very significant practical design problem. Unfortunately, guidance with respect to the influence of edge restraint and its effect in combination with end restraint is not complete and not available within the new codes.

5.3.1 Minimum reinforcement area

Previously, BS 8007 used the term critical steel ratio (ρ_{crit}) to define the area of steel required such that the reinforcement yields at the same time the concrete reaches its ultimate tensile stress. This was believed to allow a measure of control over the crack spacings and widths. Below this value, steel could potentially yield and crack widths would be non-compliant.

BS EN 1992-1-1 uses the term minimum reinforcement area, $A_{s,min}$ to define the required area of steel to negate the potential for the steel to yield once a crack has formed. Expression 7.1 of BS EN 1992-1-1 defines $A_{s,min}$ as:

$$A_{s,min}\, \sigma_s = k_c\, k f_{ct,eff}\, A_{ct} \tag{5.1}$$

where

- $A_{s,min}$ is the minimum area of reinforcing steel within the tension zone
- A_{ct} is the area of concrete within the tension zone. The tension zone is that part of the section that is calculated to be in tension just before formation of the first crack
- σ_s is the absolute value of the maximum stress permitted in the reinforcement immediately after formation of the first crack. This may be taken as the yield strength of the reinforcement, f_{yk}
- $f_{ct,eff}$ is the mean value of the tensile strength of the concrete effective at the time when the cracks may first be expected to occur: $f_{ct,eff} = f_{ctm}$. However, if cracking is expected to occur earlier than 28 days, i.e. due to short-term movements, use the 3-day value, $f_{ctm}(t)$
- k is a coefficient that allows for non-uniform self-equilibrating stresses, which leads to a reduction in restraint forces
- k_c is a coefficient that takes account of the stress distribution within the section immediately prior to cracking and of the change of the lever arm

As before in BS 8007, by equating the yield force in the steel with the tensile force in the concrete, the minimum area of reinforcing steel can be shown to be proportional to the ratio of the tensile strength of the concrete to the yield strength of the steel. For instance, by substituting f_{yk} for σ_s and $f_{ctm(t)}$ for $f_{ct,eff}$ and rearranging gives:

$$\frac{A_{s,min}}{A_{ct}} = \frac{k_c\, k f_{ctm(t)}}{f_{yk}} \tag{5.2}$$

However, whereas in BS 8007, ρ_{crit} was equal to the ratio of the 3-day tensile strength of the concrete to the yield strength of the steel, in BS EN 1992-1-1 the relationship is proportional, and includes coefficients k_c and k, which are influenced by the nature of restraint (external or internal). The conditions of restraint also influence the value of A_{ct} where it is normal to estimate the area of steel in each face (see section 5.3.4).

5.3.2 Crack spacing

The traditional theory suggests that in an end-restrained member or a member subjected to pure tension, such as an axially reinforced concrete prism, if sufficient reinforcement is provided to control the cracking, then the probable spacing of the cracks may be estimated using the ratio of bar diameter to steel ratio (ϕ/ρ). The popular theory and certainly that presented in BS EN 1992-1-1, as it was in BS 8007, is that as the short and long-term movements develop, cracks form in sequence when the bond force between the reinforcement and the concrete becomes greater than the tensile strength of the concrete. This approach is based on many investigations performed over the last 30 to 40 years both here in the UK and abroad.

According to this theory, the bond stress between concrete and the surface of the steel that accompanies the formation of a crack extends for a length equal to half the crack spacing (Figure 5.8). Equating the two forces gives

$$f_b s \sum u_s = f_{ct} bh \qquad (5.3)$$

where

$\sum u_s$ = total perimeter of bars in the width considered;
f_b = average bond stress adjacent to a crack;
s = bond length necessary to develop cracking force;
f_{ct} = tensile stress in concrete;
bh = area of concrete.

Writing steel ratio $\rho = \frac{A_s}{bh}$ (neglecting concrete area taken up by steel) and the ratio

$$\frac{\text{total perimeters}}{\text{total steel area}} = \frac{\sum u_s}{A_s} = \frac{\pi\phi \times (\text{number of bars})}{\frac{\pi}{4}\phi^2 \times (\text{number of bars})} = \frac{4}{\phi} \qquad (5.4)$$

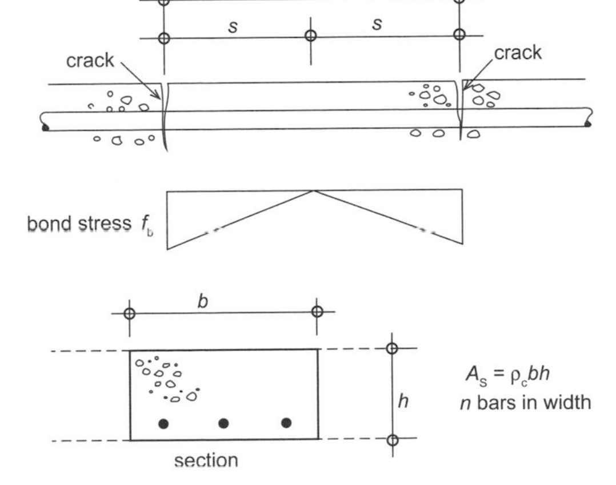

Figure 5.8 *Bond stress and crack formation.*

where ϕ = bar size (or equivalent for square or ribbed bars), substitution in (5.3) gives

$$f_b s\left(\frac{4}{\phi}A_s\right) = f_{ct}bh$$

$$\therefore \left(\frac{f_{ct}}{f_b}\right)\left(\frac{1}{\rho_c}\right)\left(\frac{\phi}{4}\right)$$

$$= \left(\frac{f_{ct}}{f_b}\right)\frac{\phi}{4\rho_c}$$

and the maximum crack spacing

$$s_{max} = 2s = \left(\frac{f_{ct}}{f_b}\right)\frac{\phi}{2\rho_c} \quad (5.5)$$

However, whereas previously (in BS 8007) the crack spacing was related only to the bond characteristics of the reinforcement and the bar diameter (Eq. (5.3) above), BS EN 1992-1-1 does now relate the spacing to a third parameter, the cover. Expression 7.11 of BS EN 1992-1-1 defines the crack spacing as:

$$S_{r,max} = 3.4c + 0.425\, \frac{k_1\, \phi}{\rho_{p,eff}} \quad (5.6)$$

where

- c is the cover to the reinforcement
- k_1 is a coefficient that takes account of the bond properties of the reinforcement. BS EN 1992-1-1 recommends a value of 0.8 for high bond bars; however, where good bond cannot be guaranteed, BS EN 1992-1-1 recommends a reduction in bond strength by a factor of 0.7, on the premise that the crack spacing is very sensitive to bond. Subsequently, k_1 should be taken as 1.14
- ϕ is the bar diameter
- $\rho_{p,eff}$ is the ratio of the area of reinforcement to the effective area of concrete ($A_s/A_{c,eff}$)
- $A_{c,eff}$ is the effective area of concrete in tension around the reinforcement to a depth of $h_{c,eff}$, where $h_{c,eff}$ is the lesser of $h/2$, $2.5(c + \phi/2)$ or $(h - x)/3$ (see Figure 3.8)

The coefficients 3.4 and 0.425 are recommendations that can be found in the UK National Annex to BS EN 1992-1-1 Note: $A_{c,eff}$ is not the same as A_{ct}, which is defined above in the section for minimum reinforcement area.

The introduction of the cover parameter is to account for (a) the shape of the crack–there is significant evidence to indicate that the crack width is not uniform (i.e. the same dimension at the surface of the steel as it is at the concrete surface–see comments in Chapter 3) but that it is in fact greater at the concrete surface than at the bar surface and (b) the shear deformation of the concrete cover between the bar and the concrete cover.

A major concern with Eq. (5.3) above (which is based purely on the bond slip theory) is that it assumes that bond failure has to have occurred at the position of a crack. The CEB Model Code 90 suggests that the maximum bond stress is achieved at a slip of 0.6 mm. The crack width at the steel surface, assuming there is equal slip either side of the crack, will be 1.2 mm. As mentioned above, the crack width at the concrete surface is now expected to be greater than at the steel surface; it is at least twice the width, potentially more. Hence, the ultimate bond stress would not be reached until the concrete surface crack exceeds a value of 2 to 3 mm, which is 10 times the actual widths that are of concern under the serviceability limit state.

Beeby has for a number of years questioned the importance and function of the (ϕ/ρ) parameter (Beeby, 2004, 2005). He has cited the work of several researchers, highlighting the fact that in many instances, where investigators thought they were isolating the (ϕ/ρ) parameter as the only single variable, this was not actually the case. (He also highlighted the inconsistencies between investigations, i.e. the variation in where and how cracks were measured, and the misleading assumption that the behaviour of the concrete in tension surrounding bars in flexural tests could be predicted from tests on pure axial tension specimens.) Often, the dimensions of the specimens, the bar spacing and the diameter of the bars varied, all important factors in influencing crack spacing and width. Where tests were performed with (ϕ/ρ) being the only variable, no relationship with crack spacing / width was evident. In fact, the analysis showed that cover is a far more important parameter in defining cracking behaviour.

The introduction of the cover parameter in BS EN 1992-1-1 is seen as a step in the right direction when it comes to estimating cracking behaviour. The influence of (ϕ/ρ) should not be ignored; however, it is believed that reliance on only this classical approach only provides a simplified version of the behaviour which will lead to errors. A more robust approach is perhaps to consider a combination of the classical approach with the cover parameter along the lines of the approach suggested in BS EN 1992-1-1, but with more emphasis on the effect of the cover (and the choice of an appropriate cover parameter) and, where appropriate, by making a clear distinction between cracking due to imposed strain and flexural cracking.

5.3.3 Crack widths

In BS 8007, the crack width, w_{max} is related to the crack spacing, s_{max} and the effective strain, ε (the total contraction strain that results from the sum of the shrinkage and thermal strain (i.e. shortening) less the average tension strain in the concrete between the cracks (i.e. lengthening)):

$$w_{max} = s_{max}\,\varepsilon \tag{5.7}$$

The effective strain, ε was derived:

$$\varepsilon = (\varepsilon_{te} + \varepsilon_{cs} - 0.5\varepsilon_{ult}) \tag{5.8}$$

where

- ε_{te} is from the 'short-term' thermal contraction from peak temperature to ambient temperature and 'long-term) seasonal changes
- ε_{cs} total shrinkage strain (drying shrinkage only)
- ε_{ult} ultimate concrete tensile strain

At the time of the introduction of BS 8007, there was insufficient information available to enable precise values for the various parameters to be given. As such, the ultimate concrete tensile strain, ε_{ult} was assumed to be 200 microstrain and the shrinkage strain in the concrete, minus creep strain, was assumed to be about 100 microstrain and therefore, in the formula above, equates with the value of $0.5\varepsilon_{ult}$. The remaining strain to be considered, ε_{te}, was due to cooling from the peak of hydration temperature T_1 to ambient temperature. There was also a further variation in temperature T_2 due to seasonal changes after the concrete in the structure had hardened.

When considering the strain due to temperature, T_1, an effective coefficient of expansion of one half of the value for mature concrete was recommended due to the high creep strain in immature concrete. For mature concrete and seasonal variations due to temperature, T_2, the tensile strength of the concrete is lower compared with the bond strength, hence s (spacing) is much less for mature concrete when T_2 is appropriate; hence the actual contraction could, it was assumed, be effectively halved. The strain equation then became:

$$w/s_{max} = 0.5\alpha\,(T_1 + T_2) + 100 - (0.5 \times 200) \qquad (5.9)$$
$$= 0.5\alpha\,(T_1 + T_2)$$

Therefore, crack width, $w = s_{max}\,0.5\alpha\,(T_1 + T_2)$
where α = coefficient of linear expansion of concrete
In BS EN 1992-1-1, the crack width, w_k is calculated using

$$w_k = S_{r,max}\,(\varepsilon_{sm} - \varepsilon_{cm}) \qquad (5.10)$$

where

$S_{r,max}$ is the maximum crack spacing
ε_{sm} is the mean strain in the reinforcement under the relevant combination of loads, including the effect of imposed deformations and taking into account the effects of tension stiffening
ε_{cm} is the mean strain in the concrete between cracks

This equation is similar in form to that presented previously in BS 8007, except BS 8007 used ε, which was the effective strain, instead of $(\varepsilon_{sm} - \varepsilon_{cm})$, which is the crack-inducing strain. The general guidance provided in BS EN 1992-1-1 is modified specifically for the design of liquid-retaining structures in BS EN 1992-3 where the crack-inducing strain is defined depending on the type of external restraint (edge or end restraint).

Restraint of a member

BS EN 1992-3 includes Figure M2, which attempts to illustrate the difference between the cracking in end and edge restraint situations. Expressions M.1 (the maximum crack width in end-restrained members)

$$(\varepsilon_{sm} - \varepsilon_{cm}) = [\sigma_s - k_t\,(f_{ct,eff}/\rho_{p,eff})(1 + \alpha_e\rho_{p,eff})]/E_s \qquad (5.11)$$

where
α_e is the modular ratio (E_s/E_c)
ρ is the reinforcement ratio

f_{ct} the tensile strength of the concrete
k_c, k are defined in Section 5.3.1

and M.3 (crack width in edge-restrained walls)

$$(\varepsilon_{sm} - \varepsilon_{cm}) = R_{ax}\varepsilon_{free} \tag{5.12}$$

where

R_{ax} is the restraint factor, and
ε_{free} is the strain that would occur if the member was completely unrestrained

are graphically presented. Figure M.2 also presents the development of cracking prior to the stabilised cracking phase in end-restrained members being achieved. In summary, the figure shows that for an end-restrained member, the crack width is constant for any number of cracks prior to a stabilised crack pattern being achieved. However, for an edge-restrained wall, once a crack has occurred due to the imposed strain, the crack width will continue to grow with further imposed deformation.

End restraint

The situation considered is one of pure tension and where the imposed deformations arise from the shortening of the member due to shrinkage and / or a change in temperature. The development of formulae for the prediction of crack widths given in clause 7.3.4 of BS EN 1992-1-1 can be arrived at by considering the description of the cracking phenomenon below (Principles of cracking); this is based on the derivation by Beeby (1990) and has been proven against experimental observations. When all crack widths have formed the crack width, w is given by a simple compatibility statement:

$$w = S_{rm}\varepsilon_m \tag{5.13}$$

where S_{rm} is the average crack spacing and ε_m is the average strain. This assumes that all the extension that occurs due to the formation of a crack is accommodated in that crack. In design, it is the maximum rather than the average crack width that is desired. Hence in BS EN 1992-1-1 the final expression (Expression 7.8) is:

$$w_k = S_{r,max}\,(\varepsilon_{sm} - \varepsilon_{cm}) \tag{5.14}$$

where the average strain, ε_m, is equal to the strain in the reinforcement, accounting for tension stiffening, ε_{sm}, less the average strain in the concrete at the surface, ε_{cm}.

Principles of cracking

The approach almost universally used to explain the basic cracking behaviour of reinforced concrete is to consider the cracking of a concrete prism reinforced with a central bar, which is subjected to pure tension. Figure 5.9 illustrates the conditions in such a prism as it is subjected to increases in tensile strain.

As the strain is increased, the stress in the concrete and steel increases until the tensile strength of the concrete is reached (Figure 5.9(b)). At this stage, the stress in the concrete is just equal to the tensile strength of the concrete, f_{ct} and the stress in the steel is $f_{ct}E_s/E_c$ or $\alpha_e f_{ct}$ where α_e is the modular ratio, E_s/E_c.

At this point the concrete will crack and there will be a considerable redistribution of forces. (In BS 8007, the theory proposed that for an end-restrained member,

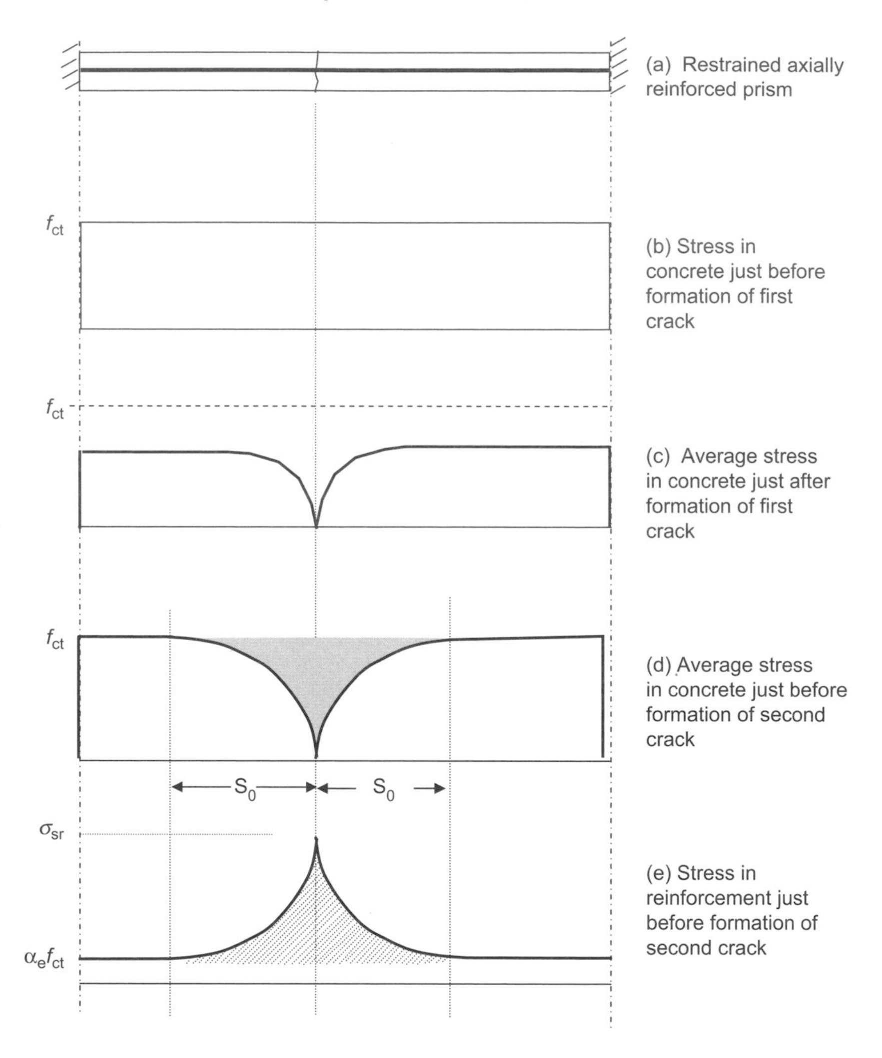

Figure 5.9 *Stress conditions in a prism during the early stages of cracking (strain control).*

all possible cracks formed at once at the cracking load. However, a more up-to-date view is that a single crack forms when the tensile stress (from the restrained strain) reaches the tensile strength of the concrete. For the restraint of a member at its end, $(\varepsilon_{sm} - \varepsilon_{cm})$ is, therefore, determined by the tensile strength of the concrete at the time at which the first crack occurs as, for this restraint condition, the magnitude of restrained strain has no effect on individual crack widths, although it may affect the number of cracks that occur.) The stress in the concrete at, or immediately adjacent to the crack, must drop to zero. However, with increasing distance from the crack, bond stresses at the bar-concrete interface will transfer force from the bar to the concrete until, at some distance, S_0, from the crack, the stress reaches a uniform value. The effect of the crack is thus to shed load from a portion of the concrete in the region of the crack.

This means that less concrete is resisting the forces induced by the strain and the stiffness of the prism reduces. If the strain remains constant, this reduction in stiffness leads to a reduction in the force and hence the stresses in the prism. As a consequence, the uniform stress away from the crack decreases to below the tensile strength of the concrete. This situation immediately after the formation of the first crack is shown in Figure 5.9(c). The overall extension, and hence the average strain, is the same in Figures 5.9(b) and (c).

Now consider some further increase in applied extension. As the average strain increases, the stress in the steel and the concrete increase until the tensile strength of the concrete is again reached. This situation is illustrated in Figure 5.9(d), which shows the distribution of the average concrete stress and (e), which shows the distribution of steel stress. These illustrate the state at the instant before the formation of the second crack. From considerations of equilibrium and compatibility we can define the stress and strains at the crack and at a section distant from the crack where the stresses are uniform. This is done as follows.

Away from the crack we know the stress in the concrete is equal to the tensile strength, f_{ct} and that the stress in the steel is $\alpha_e f_{ct}$. The force carried by the prism is thus:

$$T = f_{ct} A_c + \alpha_e f_{ct} A_s \tag{5.14}$$

where T is the tensile force to which the prism is subjected.
If the reinforcement ratio, ρ, is defined as A_s/A_c, then the equation for the force can be rewritten as:

$$T = f_{ct} (1 + \alpha_e \rho) A_c \tag{5.15}$$

The force in the bar at the crack must be equal to T since all sections of the prism must carry the same force. Consequently, the stress in the reinforcement at the crack must be:

$$\sigma_{sr} = T/A_s = \alpha_e f_{ct} (1 + 1/\alpha_e \rho) \tag{5.16}$$

By considering the deformations, an expression can now be derived for the crack width. Since both the steel and the concrete are assumed to be elastic, the strains in the steel and concrete will be proportional to the stresses shown in Figures 5.9(d) and (e). The width of the first crack just before the formation of the second crack will be equal to the extension of the steel caused by the crack, which will be proportional to the shaded area in Figure 5.9(e) plus the shortening of the concrete, which will be proportional to the shaded area in Figure 5.9(d). Any area can be expressed as the base times the perpendicular height times a constant of integration depending on the shape of the curve considered. Assuming that the shaded area in Figure 5.9(d) has the same form as the shaded area in Figure 5.9(e), we can take the same constant of integration for both. The crack width can now be written as:

$$w = \int_{+s_0}^{-s_0} (\sigma_{sr} - \alpha_e f_{ct})/E_s + \int_{+s_0}^{-s_0} f_{ct}/E_c \tag{5.16a}$$

$$= 2S_0\beta \left[(\sigma_{sr} - \alpha_e f_{ct})/E_s + f_{ct}/E_c\right] \tag{5.16b}$$

where β is a constant of integration. Substituting for σ_{sr} from (5.16) and writing $E_c = E_s/\alpha_e$ gives

$$w = 2\alpha_e f_{ct}\, S_0 \beta\, (\Sigma\mu\beta_0\lambda\, \alpha_e\, \rho)/E_s \tag{5.17}$$

From the classical theories of cracking, $S_{max} = 2S_0$. Also, there is much research data to show that β is closely approximated by a value of 0.5. Substituting for these values gives

$$w_{max} = 0.5\alpha\, f_{ct}\, S_{max}\, (1+1/\alpha_e\, \rho)/E_s \tag{5.18}$$

To gain a more general appreciation of cracking behaviour, it will first help to clarify just how crack widths are predicted to vary as strains are increased. This is illustrated schematically in Figures 5.9 and 5.10.

When the first crack forms, the stresses immediately reduce due to the reduction in stiffness caused by the crack. Since the crack width is proportional to the stress in the

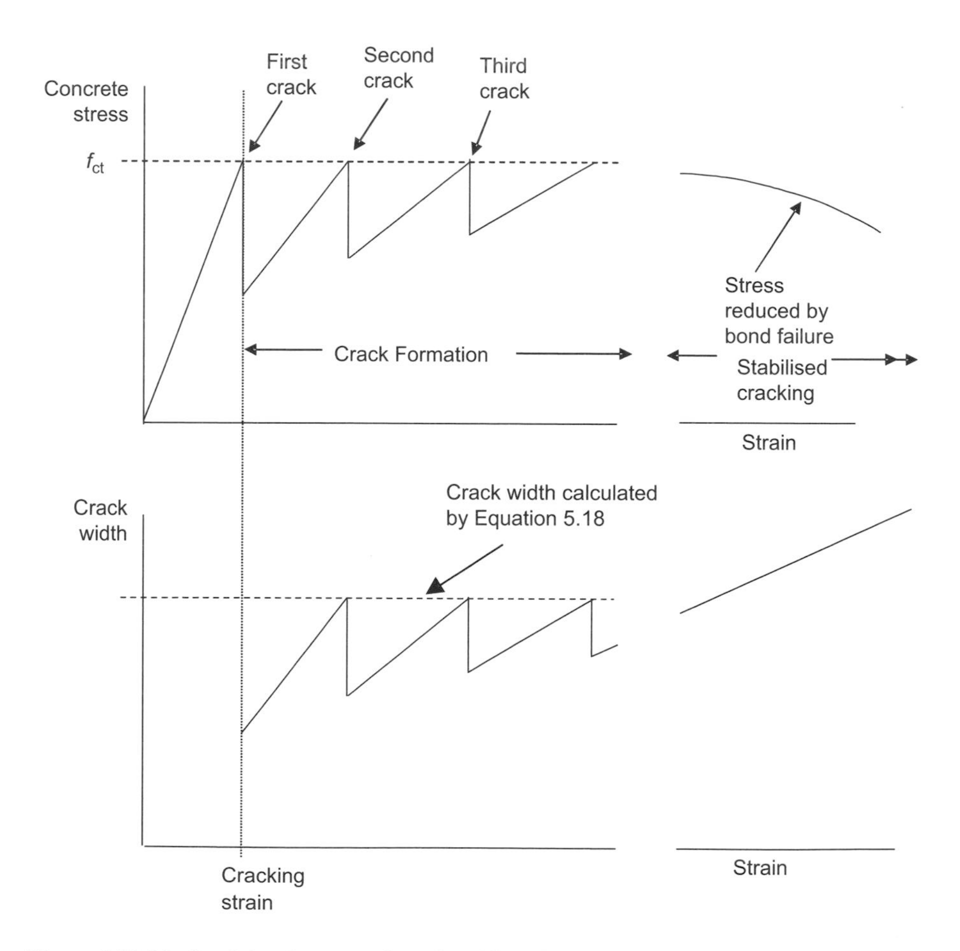

Figure 5.10 *Idealised development of crack width and concrete stress with increase in imposed strain.*

concrete, the crack width is relatively small. As the strain increases, the crack width increases until the state shown in Figures 5.9(d) and (e) is reached. At this point the crack width is as given by Eq. (5.18). As soon as the second crack forms, the stresses again reduce due to the reduction in stiffness caused by the second crack. Increase in strain causes an increase in the width of both the first and second crack until they again reach the width given by Eq. (5.18) just before the third crack forms. This process is repeated until all possible cracks have formed.

Beyond this (stabilised cracking), the cracks increase in proportion to the applied strain and are independent of the tensile strength of the concrete. The crack width–strain response may thus be seen to have two phases:

(i) the crack formation phase;
(ii) the stable cracking phase.

In general, it will be the crack formation stage that will be of importance for cracking due to end restraint as it is unlikely that the imposed strains will exceed those necessary to establish the stabilised crack pattern (in practice, the restrained strain will have to be of the order of 1000 microstrain for the stabilised cracking phase to be achieved, which is unreasonable in the majority of cases). This is an idealised picture of events. One factor that has not been taken into account is the inherent variability in the concrete from section to section.

The first crack must form at the weakest section (assuming perfectly applied pure tension to the prism). As a consequence, all other sections will be stronger and the second crack will occur at a higher stress than the first. Similarly, the third crack will form at a higher stress than the second and so on. The crack widths just before the formation of each new crack will therefore be slightly larger than the crack widths at the formation of the previous crack. The variation in the concrete strength will also lead to variations in S_0 from crack to crack so that the initial cracks will not all be the same size but will vary somewhat randomly. More variations will be introduced as more cracks form and an increasing number of the spacings are reduced to below $2S_0$.

An important point to note is that each crack has a global effect since the formation of each crack reduces the stiffness of the whole member.

Not considered above is the effect of incomplete restraint. It can be shown that a reduction in the restraint has no effect on the theoretical crack widths but will increase the imposed strain required to cause the first crack to form. Equation (5.18) is therefore not influenced by the restraint.

There are a variety of ways in which this type of restraint can be dealt with in a code. One possibility is simply to introduce Eq. (5.18). Another is to consider the equation given in BS EN 1992-1-1 for the prediction of crack width (Expression 7.8). This is:

$$w_k = S_{r,max}\,(\varepsilon_{sm} - \varepsilon_{cm}) \qquad (5.18a)$$

Inspection of Eq. (5.16) will show that Eq. (5.18a) may be used to calculate the crack width for imposed deformation in the crack formation stage if:

$$(\varepsilon_{sm} - \varepsilon_{cm}) = 0.5\alpha_e f_{ct}(1 + 1/\alpha_e\,\rho)/E_s \qquad (5.19)$$

Substituting $k_c k f_{ct,eff}$ for f_{ct} and a certain amount of algebraic manipulation will result in Equation M1 in BS EN 1992-3.

Another possibility is to write:

$$\sigma_s = k_c k f_{ct,eff}/\rho \quad (5.20)$$

The resulting value can then be used with Figures 7.103N and 7.104N of BS EN 1992-3 to obtain a suitable arrangement of reinforcement.

Edge Restraint

Whereas with end restraint situations, the crack width depends on the tensile strength of the concrete up until the stabilised crack pattern has been established, at which point it depends on the yielding of the steel, for edge-restrained situations, the crack width depends on the restrained imposed strain and not the tensile strength of the concrete; $(\varepsilon_{sm} - \varepsilon_{cm})$ is, therefore defined by the magnitude of the restrained strain.

Earlier in Section 5.1, a wall totally restrained by its base was considered (Figure 5.2 is repeated again below). This is not a serious limitation since all that is required is that, instead of the total free strain being taken into account, if the base is not fully restrained, a strain is used that is the difference between the full free strain, ε_{free}, and the shortening of the base.

When the shortening, due to either shrinkage of thermal movement or a combination of the two, becomes large enough, cracks will form. The stress in the concrete at this primary crack is zero. Figure 5.2 (repeated below) shows the situation near one such crack.

Initially, it is assumed that there is no reinforcement. With increasing distance from the primary crack (which will be the full height of the wall), stress is transferred to the wall by shear at the interface with the base until, at some distance L_o from the crack, the stress distribution is unaffected by the crack–the relief of stress caused by the crack is, therefore, purely local. This is fundamentally different to the situation where a member is restrained at its ends. In the end-restrained case, the crack reduces the stiffness of the whole system and hence reduces the stresses throughout; the effect of the crack is global. The stresses are not relieved beyond L_o from the crack and

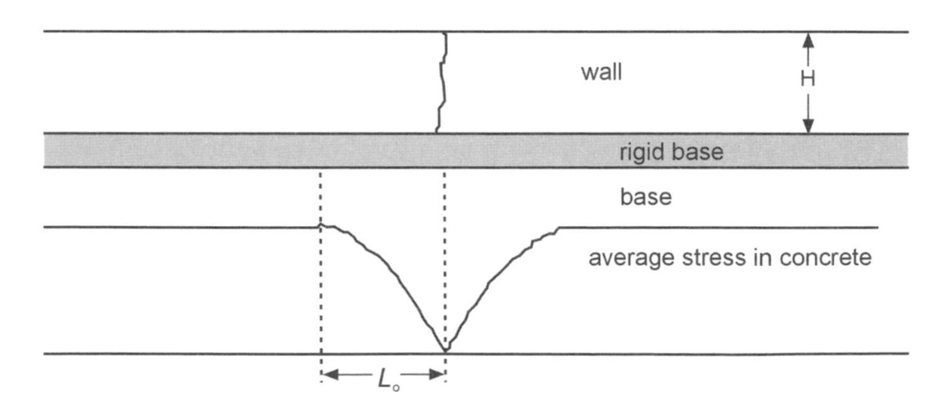

Figure 5.2 *Conditions in a base-restrained wall after initial cracking.*

further cracks may form in the unrelieved areas. These cracks will have no influence on the first crack. (Interestingly, it is therefore the restraint itself that controls the cracking, as it limits the degree to which strain relief occurs within the distance L_0.)

In the case where horizontal reinforcement is provided, the effect of reinforcement must be to improve the transfer of stress to the concrete with increasing distance from the primary crack and thus to reduce the crack width to below that obtained for the situation with no reinforcement. The effect of the reinforcement will be to provide a force across the free surface that will act to reduce the deformations to less than those calculated for zero reinforcement.

The case considered above assumed complete fixity along the bottom edge of a wall. In fact, the fixity will inevitably be less than total. This may be handled by including a restraint factor, R_{ax}, such that $R_{ax} = 1$ corresponds to total restraint and $R_{ax} = 0$ corresponds to zero restraint. Hence, expression M.3 in BS EN 1992-3:

$$(\varepsilon_{sm} - \varepsilon_{cm}) = (R_{ax})\varepsilon_{free} \quad (5.21)$$

The relationship between crack width and imposed strain will therefore appear as sketched below in Figure 5.11. The concept of a crack formation phase is meaningless for edge-restrained members.

Calculation of the crack-inducing strain

The introduction of BS EN 1992-1-1 has provided additional guidance on some of the factors that influence the free strain (i.e. autogenous and drying shrinkage). However, its recommendation of a coefficient of thermal expansion for concrete of 10με/°C is limiting. It also lacks appropriate guidance on the principle factors affecting early age thermal cracking i.e. potential temperature rises and differentials. And in BS EN 1992-3, a maximum value of 0.5 (which includes the effect of creep) has been adopted from BS 8007 as the restraint factor (general guidance on restraint factors is provided

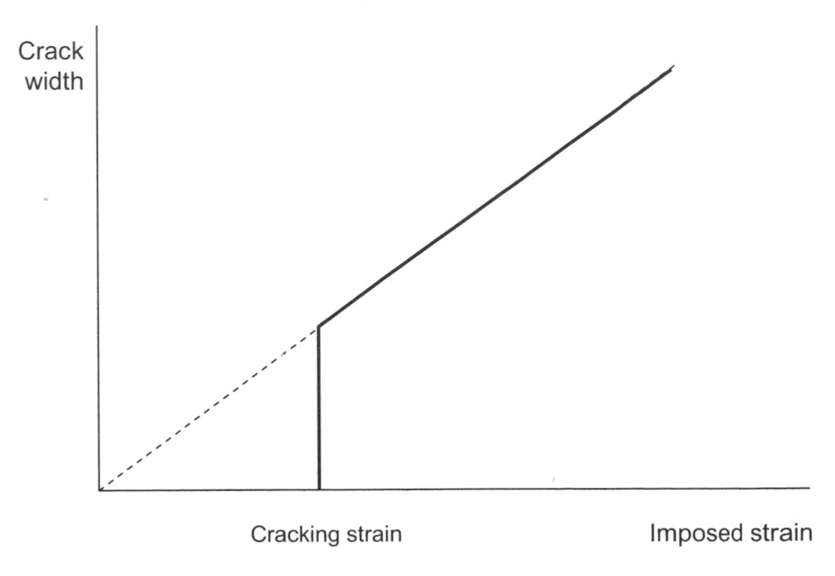

Figure 5.11 *Relation between crack width and imposed strain for edge restrained walls.*

(Annex L of BS EN 1992-3); again this has been adopted from BS 8007). With the amount of research on concrete material behaviour over the last couple of decades, it is disappointing that more was not made of this opportunity by the code. However, in 2007, a very thorough document was released by CIRIA: *Early-age Thermal Crack Control in Concrete,* which collated the current knowledge and supplemented the guidance provided in BS EN 1992-1-1. The CIRIA document proposed that the crack-inducing strain, $(\varepsilon_{sm} - \varepsilon_{cm})$ or ε_{cr} in CIRIA is equal to

$$\varepsilon_{cr} = (K_1 \{[\alpha_c . T_1 + \varepsilon_{ca}] . R_1 + [\alpha_c . T_2 . R_2] + [\varepsilon_{cd} . R_3]\}) - 0.5\, \varepsilon_{ctu}$$

Short-term movements | Long term movements

where

- T_1 is the difference between the peak temperature and the mean ambient temperature
- T_2 is the long-term fall in temperature, which takes account of the time of year at which the concrete was cast
- α_c is the coefficient of thermal expansion of concrete
- ε_{ca} is autogenous shrinkage
- ε_{cd} is drying shrinkage
- R_1 is the restraint factor that applies during the early thermal cycle
- R_2 R_3 are restraint factors applying to the long-term movements
- K_1 is a coefficient for the effect of stress relaxation due to creep under sustained loading
- ε_{ctu} is the tensile strain capacity of the concrete under sustained loading

The constant of 0.5 applied to ε_{ctu} reflects the assumption that after cracking the average residual strain in the concrete will equal half the tensile strain capacity. As can be seen from above, this assumption was also present in BS 8007. It is certainly a reasonable assumption for end-restrained conditions; however, there is currently insufficient evidence to confirm its appropriateness to edge-restrained conditions (possibly being conservative).

Within the CIRIA document, guidance is also provided for the calculation of restraint factors for different conditions (in addition to those provided in BS EN 1992-3) and in particular to the case of external edge restraint. The method for estimating edge restraint is based on an approach presented by the ACI (1990), which was updated in 2003 by Nilsson *et al.*, (2013). It is likely that the long-term restraints, R_1 and R_2, will be lower than the short-term restraint factor, R_1. Therefore, the single value of R_{ax} used in Expression M3 of BS EN 1992-3 will also be conservative.

As mentioned above, previously in BS 8007 the guidance available was less precise due to the lack of fundamental material research and there was less understanding of the conditions of restraint. The guidance provided by the CIRIA document certainly improves the assessment of free movement, and the comments in BS EN 1992-3 highlight to the designer the presence of the different restraint conditions. The approach provided in BS EN 1992-1-1 and BS EN 1992-3 is fundamentally similar to that in BS 8007 for end restraint conditions but potentially more accurate on a like-for-like basis. However, due to the lack of investigation into the effect of edge restraint on the behaviour of walls either in isolation or in combination with end restraint, the designer is still unable to quantify the complete effects of restraint on all structural elements (particularly, the steel percentage required to resist non-structural cracking). The design example in Chapter 6 includes a calculation to assess the thermal and shrinkage effects on the cracking of a wall in a buried service reservoir. The calculation follows the updated guidance made available by Bamforth in the CIRIA C660 document for a wall restrained at its base (i.e. edge restrained) and does not consider the combined effects of the two types of restraint (the wall in the example will in fact also be restrained at its ends).

5.3.4 Surface zones

Section 5.3.1 provides details of the code guidance for calculating the minimum area of steel, $A_{s,min}$, and, as shown, in order to calculate this reinforcement area, a surface zone of influence should be defined. The minimum area of steel is that area that should at least be provided to control crack spacing and width resulting from the effects of imposed deformations (strains from temperature change and shrinkage). In BS 8007, the surface zones were defined as shown in Figure 5.12, i.e. having calculated the required steel ratio, the ratio could be converted to the reinforcement area by considering two surface zones in the slab. Each surface zone is of thickness equal to one half the overall width / depth of the wall / slab, but with a maximum value of 250 mm. The calculated reinforcement for each surface zone was then required to be

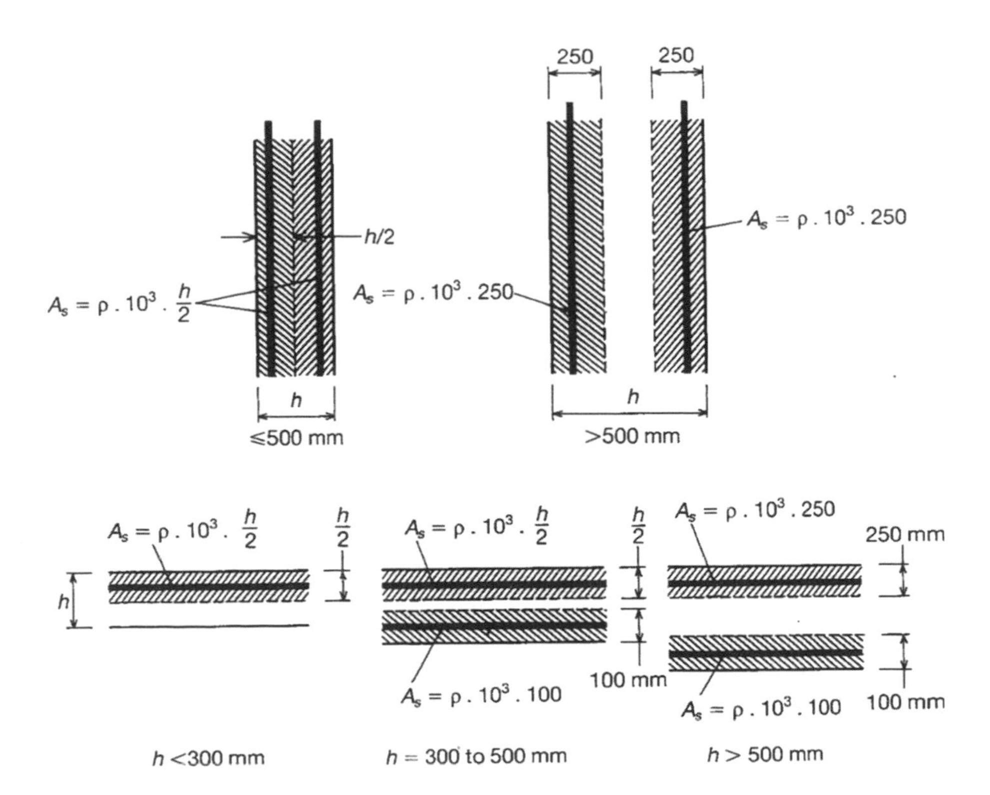

Figure 5.12 *(a) Surface zones in walls and suspended slabs (b) Surface zones in ground slabs.*

placed adjacent to that face. This approach is broadly adopted in BS EN 1992-1-1, where the coefficients k_c and k, when applied to A_{ct}, effectively produce surface zones equivalent to those in BS 8007, up to section thicknesses of 800 mm. The value of $k_c = 1.0$ for pure tension (for tension across the whole section, external restraint must be governing). Depending on the section thickness, $k = 1.0$ ($h < 300$ mm) or 0.65 ($h > 800$ mm) – interpolation is allowed between these limiting values. The coefficients k_c and k are, in effect, reduction factors, which reduce the area of concrete in tension. As mentioned before, k_c adjusts for different forms of stress distribution and can vary between a value of 0.4, which would be the case for an element in pure flexure, to a value of 1.0.

However, in BS 8007, as the maximum thickness of surface zone was 250 mm, it follows that for a wall/slab element over 500 mm in thickness, the thermal steel in each face would remain constant. This is not the case in BS EN 1992-1-1 where, for instance, for a section (in pure tension i.e. $k = 1.0$) that is 1000 mm thick, the surface zone would be 325 mm (i.e. $1000 \times 0.65 = 650$ mm; $650/2 = 325$ mm). This, therefore, means that BS EN 1992-1-1 requires additional steel to that which would have been recommended by BS 8007. It follows, therefore, that either BS EN 1992-1-1 is specifying excessive steel areas for sections greater than 800 mm or that in the past when structures have been designed to BS 8007, crack widths greater than those predicted using the code have occurred. The authors have witnessed such non-compliance in existing structures designed using BS 8007.

The fundamental approach taken in BS 8007 is very focused on the type of imposed deformation (from temperature change; autogenous shrinkage was ignored) and how this deformation is restrained. In an element where this early thermal deformation is externally restrained, the following theory was proposed to support the approach of BS 8007. Consider a slab restrained at its ends (end restraint). When cast, and during hydration as the temperature rises to peak temperature, a differential temperature profile is expected to exist, where the middle of the section is hotter than the edges of the slab. Hence the middle would expand more but is restrained externally so goes into compression; for equilibrium, the edges would go into tension. During this period where the concrete is plastic / in a pre-set condition, creep potential is high and the compressive forces in the middle are expected to be relieved by creep. (Interestingly, no mention of the tension in the edges being relieved is made; in practice, the tension developed should easily be relieved. Research has shown that at low stresses, tensile creep can be up to 10 times greater than compressive creep.) The theory suggests that, at this stage, tensile cracks may have already formed in the concrete surface during this heating stage. Once peak temperature has been reached and the concrete begins to cool, the proposed theory then goes on to suggest that the edge concrete will want to cool more rapidly than the middle region. However, in trying to do so, the middle region restrains its movement causing further tension in the concrete surface regions and therefore additional cracking. From this proposed theory, it can clearly be seen why the surface zone approach presented in BS 8007 exists and is applicable when cracks are likely to initiate at the surface (where part of the cross-section is in tension).

The Designers' Guide to BS EN 1992-1-1 does mention shrinkage (in addition to the imposed deformation from temperature change) of an externally restrained element and explains how–because a non-linear distribution of strain will exist across the section, i.e. the surface will want to shrink more rapidly than the middle of the section–again the middle region restrains the movement of the surface region leading to greater tension in the surface region and hence further cracking in this region.

The development of tension and compression described above is realistic; however, there exist two potential variations to the pattern described. Firstly, as mentioned above, it is believed that there is significant likelihood that creep would relieve any tension that developed on heating in the surface regions, negating any tension cracks forming during this period. Secondly, and more importantly, there is a significant possibility that the contraction of the middle portion of the element is greater and more rapid than the edge regions. The result of this is that the edge regions would restrain the movement of the middle regions, putting the middle regions into tension and the edge regions into compression. This would therefore lead to cracks forming in the middle of the section first. These cracks may spread to the surface regions due to continued temperature changes alone. However, the cracks most definitely could extend to the surface if we include the effect of shrinkage deformation, which as mentioned above leads to tension being developed in the surface region (and a possible reduction in the cracks in the middle section). It is interesting to note that if the tension cracks caused on heating were not relieved by creep as suggested by the approach in BS 8007, these cracks would want to close as tension is developed in the middle due to greater and rapid cooling, but also open as tension is developed in the surface due to shrinkage deformations. Ultimately, due to these variations there is the potential that the whole cross section could be in tension, not just the surface zones (or that only the middle region could be cracked). The approach presented by BS EN 1992-1-1 is therefore not considered by the authors to be completely reliable.

5.4 Joints

5.4.1 Construction joints

It is rarely possible to build a reinforced concrete structure in one piece. It is therefore necessary to design and locate joints that allow the contractor to construct the elements of the structure in convenient sections. In normal structures, the position of the construction joints is specified in general terms by the designer, and the contractor is allowed to decide on the number of joints and their precise location subject to final approval by the designer.

In liquid-retaining structures this approach is not satisfactory. The design of the structure against early thermal movement and shrinkage is closely allied to the frequency and spacing of all types of joints, and it is essential for the designer to specify on the drawings exactly where construction joints will be located. Construction joints should be specified where convenient breaks in placing concrete are required. Concrete is placed separately on either side of a construction joint, but the reinforcement is continuous through the joint. At a horizontal construction joint, the free surface of the concrete must be finished to a compacted level surface. At the junction between a base slab and a wall, it is convenient to provide a short 'kicker', which enables the formwork for the walls to be placed accurately and easily. A vertical joint is made with formwork. Details are shown in Figure 5.13.

Construction joints are not intended to accommodate movement across the joint but, due to the discontinuity of the concrete, some slight shrinkage may occur. This is reduced by proper preparation of the face of the first-placed section of concrete to encourage adhesion between the two concrete faces. Joint preparation consists of removing the surface laitance from the concrete without disturbing the particles of aggregate. It is preferable to carry out this treatment when the concrete is at least 5 days old, either by sandblasting or by scabbling with a small air tool. The use of retarders painted on the formwork is not recommended, because of the possibility of contamination of the reinforcement passing through the end formwork. The face of a construction joint is flat and should not be constructed with a rebate. It is found that the shoulders of a rebate are difficult to fill with compacted concrete, and are also liable to be cracked when the formwork is removed. Any shear forces can be transmitted across the joint through the reinforcement. If a construction joint has been properly prepared and constructed, it will retain liquid without a waterstop. Extra protection may be provided by sealing the surface as shown in Figure 5.14.

Designers are under some pressure to use waterstops in construction joints for obvious commercial reasons and also because it is thought that there is less responsibility thrown onto the designer if a waterstop is specified than if it is omitted. The author knows of instances where waterstops have been used and leaks have been widespread. In other cases, both waterstops and an external membrane have been

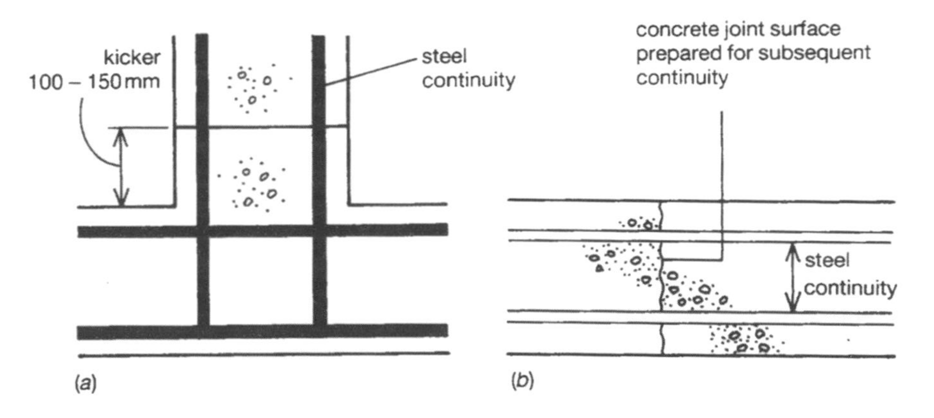

Figure 5.13 *Construction joints (a) Horizontal joint between base slab and wall (b) Vertical joint.*

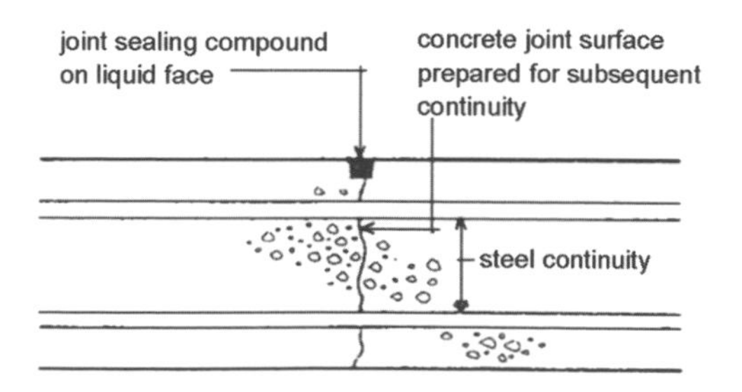

Figure 5.14 *Construction joint sealed on the liquid face.*

specified and again with completely unsatisfactory results. These examples suggest that workmanship is critical and that whatever specification is used this point is valid.

The designer must try to convince the contractor's site operatives that the work they will be executing is of the greatest importance for the correct functioning of the completed structure. Where the contractor has taken the job at a particularly low price, this may be difficult.

It is perhaps also worth stating the obvious, that it is much cheaper to spend a little more time initially to make a satisfactory job than to have to make repairs later.

5.4.2 Movement joints

Movement joints are designed to provide a break in the continuity of an element, so that relative movement may occur across the joint in the longitudinal direction. Effectively, they act as a crack; they, therefore, have the effect of reducing restraint, particularly in edge-restrained walls. The guidance in BS EN 1992-3 is less extensive than that found in BS 8007. In some way, it is also contradictory. It offers two options for control: (a) design for full restraint with no movement joints and (b) design for free movement to achieve minimum restraint (i.e. a series of closely spaced movement joints with centres at 1.5 times the wall height or 5 m, whichever is the greatest). However, it then states in part 101 b of Annex N that 'a moderate amount of reinforcement is provided sufficient to transmit any movements to the adjacent joint'. This comment infers that partial contraction joints (see 'Contraction joints' below for description) are actually acceptable under option (b).

Movement joints may provide for the two faces to move apart (contraction joints) or, if an initial gap is created, to move towards each other (expansion joints). Contraction joints are further divided into *complete contraction joints* and *partial contraction joints.*

Other types of movement joints are needed at the junction of a wall and roof slab, commonly known as a sliding joint. A typical joint of this type is shown in Figure 5.15 (for clarity, none of the potential ancillary components of such a joint i.e. unbonded / bonded membrane, free-draining material, geotextile membrane, etc. have been shown). Figure 5.16 illustrates where a free joint is required, to allow sliding to take place at the foot of the wall of a circular prestressed tank.

Contraction joints. Complete contraction joints have discontinuity of both steel and concrete across the joint (Figure 5.17), but partial contraction joints have some

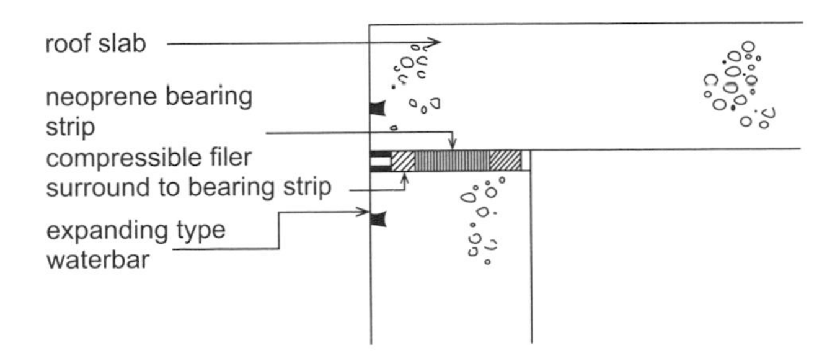

Figure 5.15 *Detail for movement joint between wall and roof slab.*

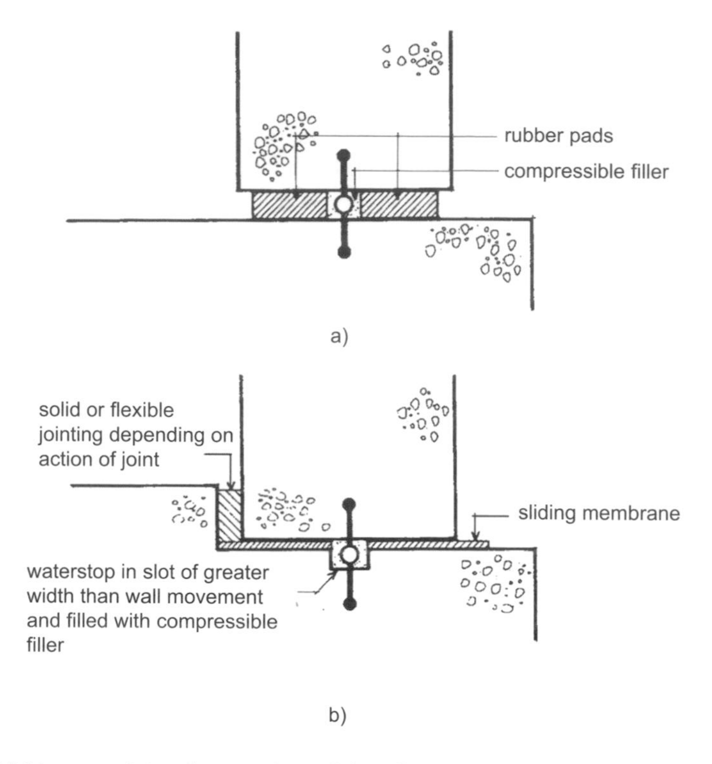

Figure 5.16 *Movement joints between base slab and wall of prestressed concrete tank (a) Rubber pad (b) Sliding membrane.*

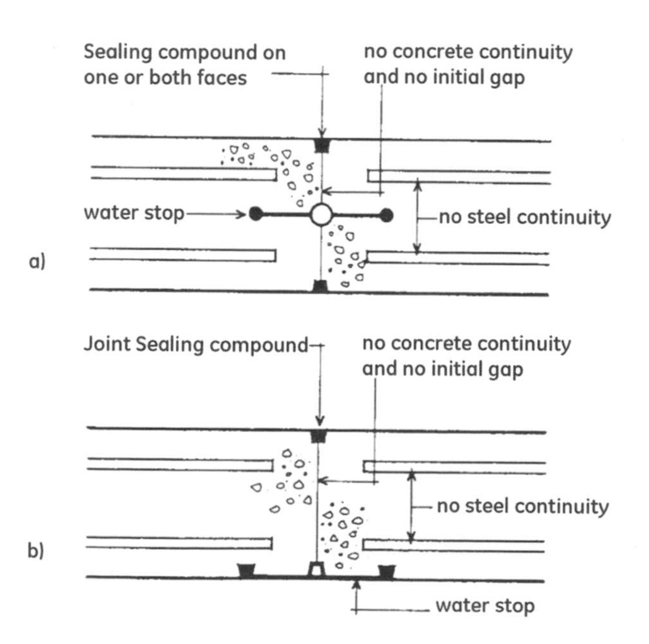

Figure 5.17 *Complete contraction joints (a) Wall (b) Floor.*

continuity of reinforcement (Figure 5.18). According to BS EN 1992-3, partial contraction joints should only have 'a moderate amount' of reinforcement continuing across the joint; this discounts the 100% continuity acceptable under BS 8007 and from interpretation by the authors would suggest a maximum of only 50% of the steel being allowed to continue across the joint, the remaining 50% being stopped short of the joint plane. It is recommended that at least a minimum area of steel equal to $A_{s,\,min}$ be continued across the joint. In determining the theoretical restraint factor of a wall with full continuity of reinforcement, CIRIA C660 advises that the existence of the partial contraction joints be ignored and that the full distance between contraction joints be used to determine the length to height ratio.

Contraction joints may be constructed as such or may be induced by providing a plane of weakness, which causes a crack to form on a preferred line. In this case, the concrete is placed continuously across the joint position, and the action of a device that is inserted across the section, to reduce the depth of concrete locally, causes a crack to form. The formation of the crack releases the stresses in the adjacent concrete, and the joint then acts as a normal contraction joint. A typical detail is shown in Figure 5.19. Great care is necessary to position the crack inducers on the same line, as otherwise the crack may form away from the intended position. Similar details may be used in walls with a circular-section rubber tube placed vertically on the joint-line on the wall centre-line, causing the crack to form.

Expansion joints. Expansion joints are formed with a compressible layer of material between the faces of the joint. The material must be chosen to be durable in wet conditions, non-toxic (for potable water construction), and have the necessary properties to be able to compress by the required amount and to subsequently recover its original thickness. An expansion joint always needs sealing to prevent leakage of

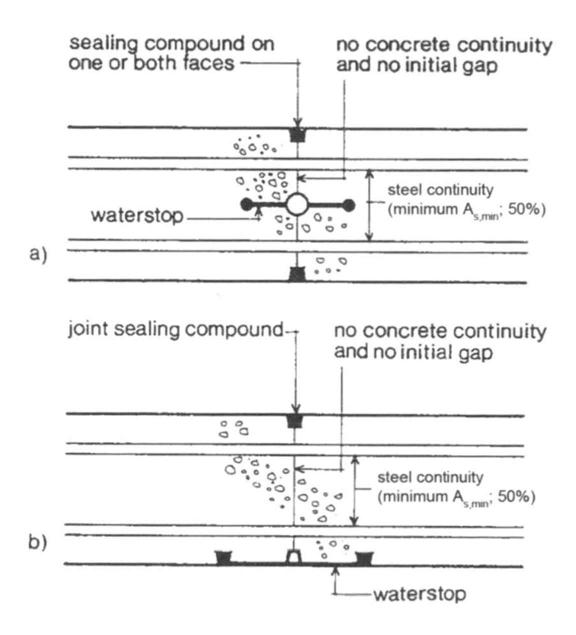

Figure 5.18 *Partial contraction joints (a) Wall (b) Floor.*

liquid. BS EN 1992-3 is very keen to advise about the incompatibility between the life of proprietary sealants and that of the structure itself. Joints therefore need to be 'inspectable and repairable or renewable'. In a wall, a water-bar is necessary, containing a bulb near to the centre, which will allow movement to take place without tearing (Figure 5.20). The joint also requires surface sealing to prevent the ingress of solid particles. By definition, it is not possible to transmit longitudinal structural forces across an expansion joint, but the designer may wish to provide for shear forces to be carried across the joint, or to prevent the slabs on each side of the joint moving

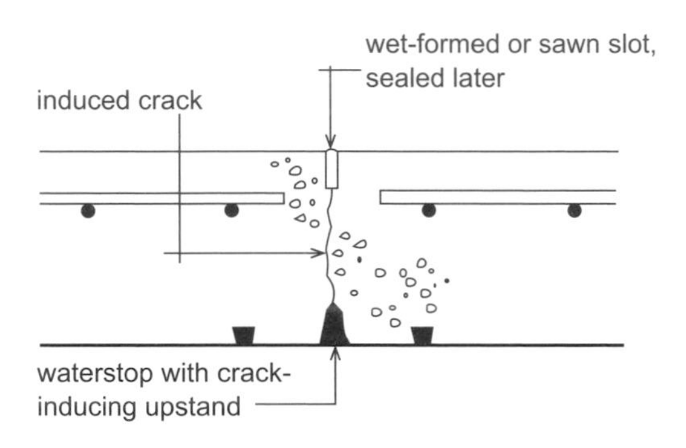

Figure 5.19 *Induced contraction joint in floor.*

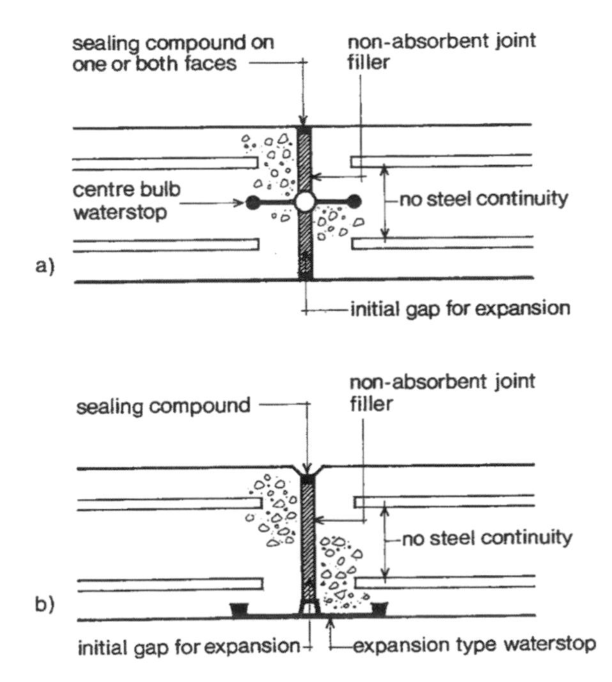

Figure 5.20 *Expansion joints (a) Floor (b) Wall.*

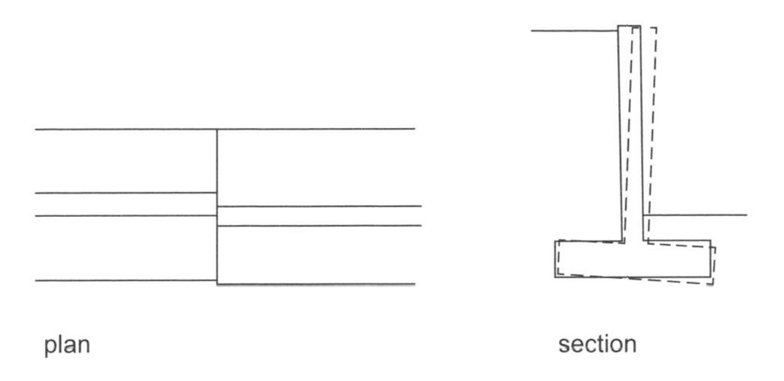

Figure 5.21 *Lateral movement at unrestrained expansion joint.*

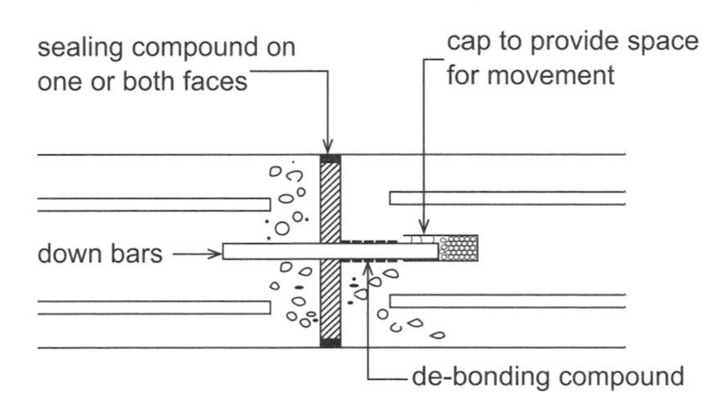

Figure 5.22 *Expansion joint including dowel bars to prevent lateral movement.*

independently in a lateral direction. If a reservoir wall and footing is founded on ground that is somewhat plastic, the sections of wall on either side of an expansion joint may rotate under load by differing amounts. This action creates an objectionable appearance and may tear the jointing materials (Figure 5.21). The slabs on either side of an expansion joint may be prevented from relative lateral movement by providing dowel bars with provision for longitudinal movement (a similar arrangement to a road slab). The dowel bars must be located accurately in-line (otherwise the joint will not move freely), be provided with an end cap to allow movement, and be coated on one side of the joint with a de-bonding compound to allow longitudinal movement to take place (Figure 5.22).

Chapter 6
Design calculations

In this chapter, basic example design calculations for a simple water-retaining structure are presented.

These hand-based calculations, made with reference to published tables and charts, illustrate the practical application of BS EN 1992-3 and other relevant Eurocodes in a realistic design situation. It is recognised that in modern structural engineering practice, computers will be used for detailed structural analysis and element design checks. However, the manual calculation methods presented here will be useful in preliminary design and also in understanding the practical application of code requirements.

6.1 Design of pump house

6.1.1 Introduction

A pump house is to be built as part of a sewerage scheme to house three electric pumps underground. The layout of the structure is shown in Figure 6.1.

The pump house comprises a buried concrete box structure with two compartments, a 'wet well' and a 'dry well' pump room. The pump motors and control equipment are housed in a building of lighter weight construction at ground level. The buried concrete box structure is required to be designed as a liquid-retaining structure.

The following aspects are considered in the calculations:

i. Concrete mix composition and concrete cover.
ii. Stability against hydraulic uplift.
iii. Lateral earth and groundwater pressures on external walls.
iv. Hydrostatic water pressures in 'wet well'.
v. Bending moments in walls.
vi. Design actions on base slab.
vii. Bending moments in base slab.
viii. SLS and ULS design checks for sample wall and base slab panels.
ix. Early-age and long-term thermal and shrinkage effects.

6.1.2 Key assumptions

The following assumptions are made:

- A geotechnical site investigation has found a dense sandy soil with no groundwater present. However, historic data for the surrounding area indicates that in extreme conditions groundwater can rise to a level of 1.0 m below ground level.

- The sequence of construction activities for the buried box structure will be as shown in Figure 6.2. Of particular significance to the design (i) all of the structural concrete work will be completed before water testing of the wet well and (ii) water testing will be carried out before backfilling to the structure.
  (Note: When undertaking the design of a water-retaining structure it is important to understand the construction and testing sequence to be used on site in order that the critical structure and loading conditions are identified.)
- The buried concrete box will be designed as a monolithic structure. For the size of structure concerned, the omission of movement joints is practical. Construction joints will be shown on drawings but water bars will not be detailed. Concrete faces forming joints with adjacent pours will require suitable preparation before casting of the next part of the structure.

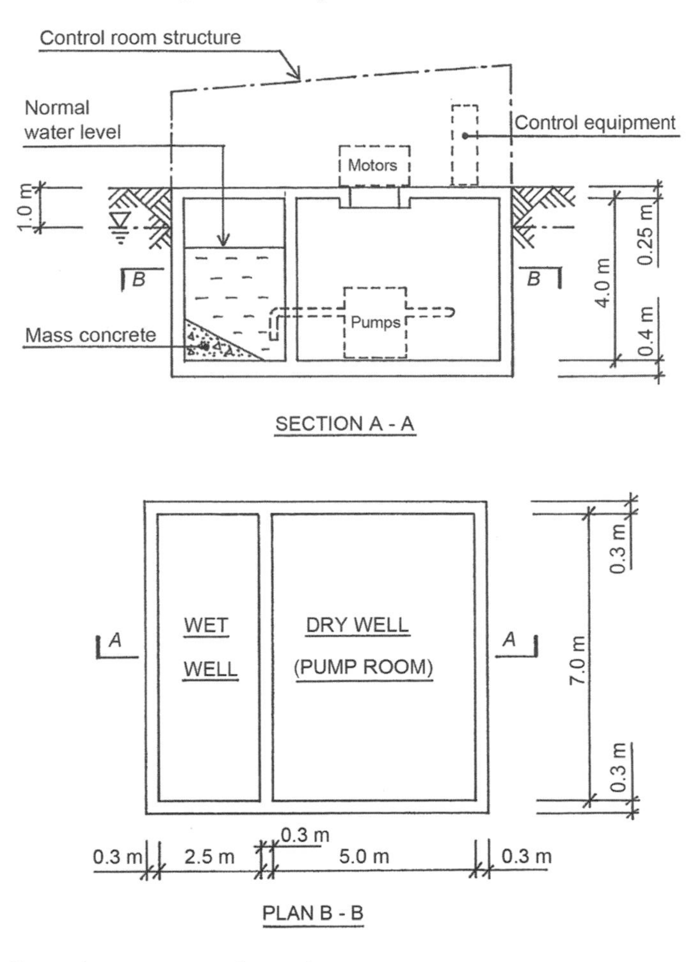

Figure 6.1 *General arrangement of pump house.*

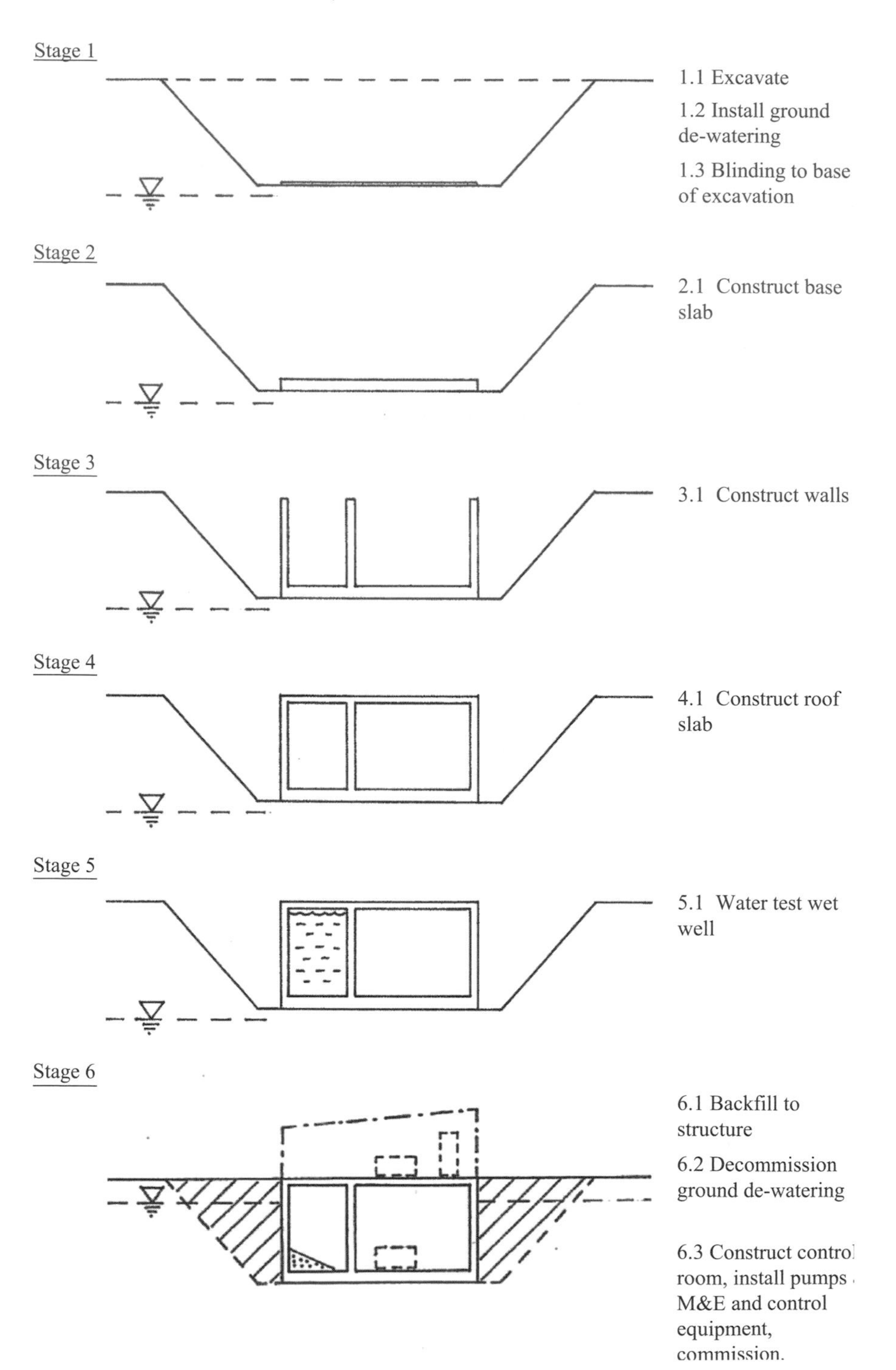

Figure 6.2 *Construction sequence of pump house.*

6.1.3 Limitations of design approach

The pump house in this example is of monolithic construction. In particular, there is full structural continuity between the perimeter walls and the roof slab. For larger structures the use of this structural form requires detailed consideration of the load effects caused by the restraint of thermal movements and creep effects. More detailed understanding of the phenomena involved has been the subject of recent research, including monitoring of movements and stresses in operational structures.

Traditional practice has been to design horizontal sliding joints at the wall-roof junctions, thereby allowing a degree of movement to the roof slab under thermal loading. However, from experience some clients now consider that that this detail gives a potential source of leakage with resultant undesirable maintenance problems over time, such as the re-sealing of joints. This applies equally whether the structure is part of a wastewater scheme, where outward leakage will cause pollution, or a drinking water scheme, where inward leakage will contaminate treated water.

The calculations presented in this example are not definitive in terms of the actions considered as, for simplicity, thermal loads (cyclic) and soil/structure interaction have not been included. These are usually analysed using computer methods. The actions that should be taken into account when designing for these effects are given in the relevant parts of the Eurocodes.

6.1.4 Calculation sheets

1. CONCRETE MIX COMPOSITION & CONCRETE COVER

BS 8500-1

Determine the minimum concrete quality and concrete cover to reinforcement for the buried box structure to meet durability requirements, in accordance with BS 8500-1.

(Note: BS 8500 is the complementary British Standard to BS EN 206-1 and contains additional provisions for use in the UK. The use of BS 8500 is required by the UK National Annex to BS EN 1992-1-1.)

Basic design assumptions

The following requirements for the design are assumed:

Intended working life	50 years
Exposure classes	XC3/4 (carbonation), XF1 (freeze-thaw)
Design ground conditions	ACEC-class AC-2
Minimum cover	30 mm
Maximum aggregate size	20 mm

Concrete quality

BS 8500-1 Table A.4

For 50-years working life + XC3/4 exposure conditions

Minimum concrete quality required is

Strength class	C28/35
Maximum w/c ratio	0.60
Minimum cement content	280 kg/m^3
Permissible cement types	All in BS 8500-1, Table 6 (except CEM IVB-V)

BS 8500-1 Table A.8

For XF/1 exposure conditions + C28/35 strength class

Limiting values of concrete composition for XF1 exposure,

Minimum air content	No requirement
Minimum cement content	280 kg/m^3
Permissible cement types	All in BS 8500-1, Table 6 (except CEM IVB-V)

BS 8500-1 Table A.9

Design chemical class (hydraulic gradient due to groundwater < 5)

DC-class	DC-2 (with no Additional Protective Measures)

BS 8500-1 Table A.11

Limiting composition and properties of concrete DC-2

Maximum w/c ratio	0.55
Minimum cement content	320 kg/m^3 (for 20 mm aggregate)
Permissible cement types	All in BS8500-1, Table 6 (except CEM IVB-V)

Concrete mix composition

Taking into account the requirements for concrete quality determined above, the composition of the concrete mix shall be:

Strength class	**C28/35**
Maximum w/c ratio	**0.35**
Minimum cement content	**320 kg/m³**
Max. aggregate size	**20 mm**
Cement types	**CIIB-V + SR** (PC + 25 – 35% PFA)
or	**CIIIA + SR** (PC + 36 – 65% GGBS)
or	**CIIIB + SR** (PC + 66 – 80% GGBS)*
Nominal concrete cover	**30 + Δc**
Maximum aggregate size	**20 mm**

(Note. In current practice cement types CIIB-V + SR and CIIIA + SR are more commonly specified than CIIIB + SR.)*

Concrete cover

BS EN 1992-1-1 §4.4.1.1

Nominal cover $c_{nom} = c_{min} + \Delta c_{dev}$

BS EN 1992-1-1 §4.4.1.2

Minimum cover for environmental conditions (for C28/35 concrete mix described above)

$$c_{min,dur} = 30 \text{ mm}$$

BS EN 1992-1-1 §4.4.1.3(1)

Allowance for deviation (i.e. reinforcement fixing tolerance)

$$\Delta c_{dev} = 10 \text{ mm}$$

BS EN 1992-1-1 §4.4.1.3(4)

For concrete cast against sand blinding $c_{min} = 40$ mm

Nominal cover for design verifications

Concrete cast against formwork

$$c_{nom} = 30 + 10 = \underline{\mathbf{40\ mm}}$$

Concrete cast against sand blinding

$$c_{nom} = 40 + 10 = \underline{\mathbf{50\ mm}}$$

2. STABILITY AGAINST HYDRAULIC UPLIFT

BS EN 1997-1 §10.2

Check uplift limit state (UPL) for buoyancy/flotation in accordance with BS EN 1997-1.

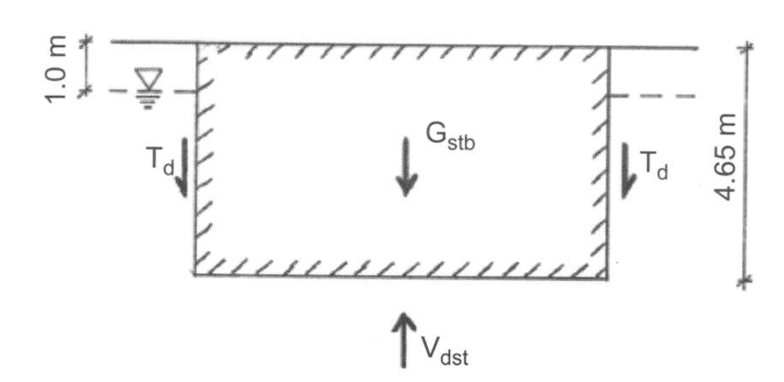

BS EN 1997-1 Eqn. 2.8

Verify stability against hydraulic uplift using BS EN 1997-1, §2.4.7.4, Equation 2.8:

$V_{dst;d} \leq G_{stb;d} + R_d$

$V_{dst;d}$ = hydraulic uplift force

$G_{stb;d}$ = self-weight of structure

R_d = additional resistance to uplift

Structure self-weight

BS EN 1991-1-1 Table A.1

Unit weight of reinforced concrete (normal % rebar)

= 24 + 1 = 25 kN/m^3

Overall dimensions of concrete box structure (ref. Figure 6.1).

8.4 m × 7.6 m × 4.65 m

Weight of reinforced concrete box structure (characteristic)

Base	25 × (8.4 × 7.6 × 0.4)	=	638
Walls	2 × 25 × (8.4 × 4.0 × 0.3)	=	504
	3 × 25 × (7.0 × 4.0 × 0.3)	=	630
Roof	25 × (8.4 × 7.6 × 0.25)	=	399
	$G_{stb;k}$	=	2171 kN

Hydraulic uplift

Unit weight of groundwater, γ_{gw} = 10 kN/m^3

For water table 1.0 m below ground level, hydraulic uplift force (characteristic)

$V_{dst;k}$ = 10.0 × (8.4 × 7.6 × 3.65) = 2330 kN

Additional resistance to uplift

Additional resistance to uplift is given by friction between the soils and the walls, T_d

$R_d = T_d$

Verification of stability against Uplift Limit State (UPL)

$V_{dst;d} \leq G_{stb;d} + R_d$

Use calculated values of structure self-weight and hydraulic uplift force to determine the minimum value of R_d required in order that stability is achieved.

BS EN 1997-1 Table A.NA.15

Partial factors on Actions for UPL limit state

$\gamma_{G;dst} = 1.1$

$\gamma_{G;stb} = 0.9$

$R_d \geq V_{dst;d} - G_{stb;d}$

$T_d \geq 1.1 \times 2330 - 0.9 \times 2171$

$\geq 2563 - 1954 = 609$ kN

To determine minimum friction (δ_d) required between soil and walls in order to generate sufficient friction force (T_d) to achieve stability, must first determine lateral earth pressure distribution on walls.

Characteristic soil properties (assumed)

$\gamma = 18$ kN/m^3

$\varphi'_k = 30°$

BS EN 1997-1 Table A.NA.16

To obtain design soil properties, use partial factors on soil parameters for UPL limit state

$\gamma_{\varphi'} = 1.25$

$\varphi'_d = \tan^{-1} [\tan \varphi'_k / \gamma_{\varphi'}]$

$= \tan^{-1} [\tan 30° / 1.25] = 24.7°$

BS EN 1997-1 Eqn 9.1

Coefficient of Lateral Earth Pressure 'At Rest'

$K_0 = 1 - \sin \varphi'_d$

$= 1 - \sin 24.7 = 0.58$

Lateral Earth Pressures against walls

At ground level (GL), $\sigma_{h\ soil} = 0$

At 1.0 m below GL

$\sigma_{h\ soil} = 0.58 \times 18 \times 1.00$

$= 10.4$ kPa

At 4.65 m below GL,

$\sigma_{h\ soil} = 10.4 + 0.58 \times (18 - 10) \times 3.65$

$= 27.3$ kPa

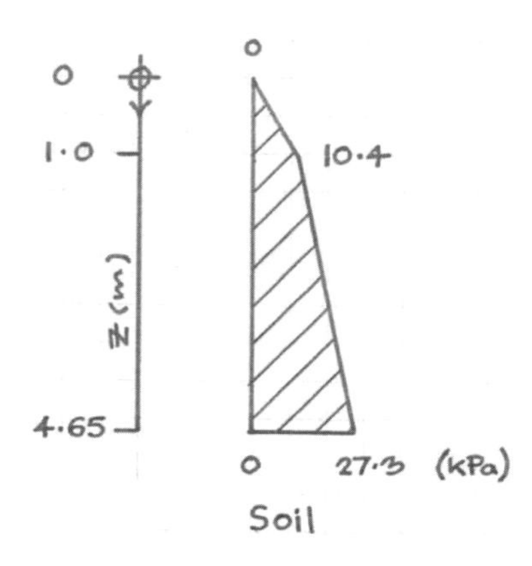

Earth pressure force acting normal to walls

$F = 0.5 \times (0 + 10.4) \times [2 \times (8.4 + 7.6) \times 1.00]$
$+ 0.5 \times (10.4 + 27.3) \times [2 \times (8.4 + 7.6) \times 3.65]$

$= 166 + 2202$

$= 2368$ kN

To achieve stability against UPL

$\delta_d \geq \tan^{-1} [T_{d,min} / F]$

$\geq \tan^{-1} [609 / 2368]$

$= 14.4°$

BS EN 1997-1 §9.5.1(6)

$\delta_d = k \times \varphi$ with limiting value of k is 2/3(= 0.667)

$k = \delta_d / \varphi$

$= 14.4° / 24.7°$

$= 0.58 <$ limiting value

So, adequate friction force, T_d, can be generated in order to stabilise the buried structure against hydraulic uplift.

Structure is stable at UPL limit state

3. LATERAL EARTH & GROUNDWATER PRESSURES ON EXTERNAL WALLS

BS EN 1997-1

Determine lateral earth and groundwater pressures on external walls of buried box structure for Ultimate Limit State (STR) and Serviceability Limit State (SLS) in accordance with BS EN 1997-1.

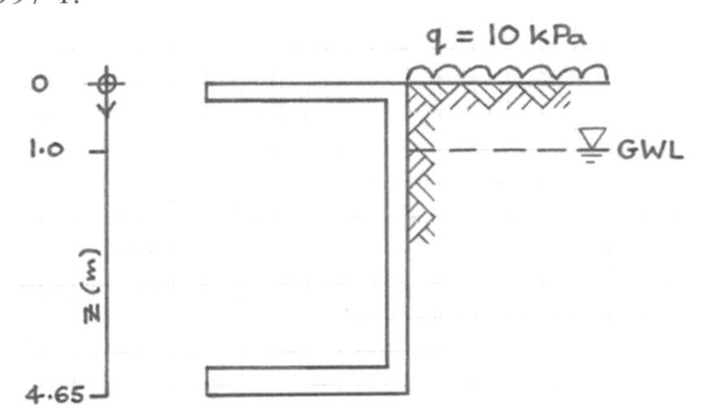

Basic data

The following data are assumed for design:

Soil properties	γ	=	18 kN/m³
	φ'_k	=	30°
	δ	=	0°
Groundwater	γ_{gw}	=	10 kN/m³
Surcharge Live Load	q	=	10 kPa

Ultimate Limit State (STR)

Assume EC7 'Combination 1' earth pressures will govern the design of structural elements (A1 "+" M1 "+" R1)

BS EN 1997-1 Table A.NA.3
BS EN 1990 Table NA.A1.2(B)

Use partial factors on actions Set A1

γ_{Gsup} = 1.35

γ_{Ginf} = 1.00

γ_Q = 1.50

BS EN 1997-1 Table A.NA.4

Use partial factors on soil parameters Set M1

$\gamma_{\varphi'}$ = 1.00

Design soil parameters

φ'_d = $\tan^{-1}[\tan\varphi'_k / \gamma_{\varphi'}]$

= $\tan^{-1}[\tan 30° / 1.00]$ = 30°

BS EN 1997-1 Eqn 9.1

Coefficient of lateral earth pressure 'at rest'

K_0 = $1 - \sin\varphi'_d$

= $1 - \sin 30°$ = 0.50

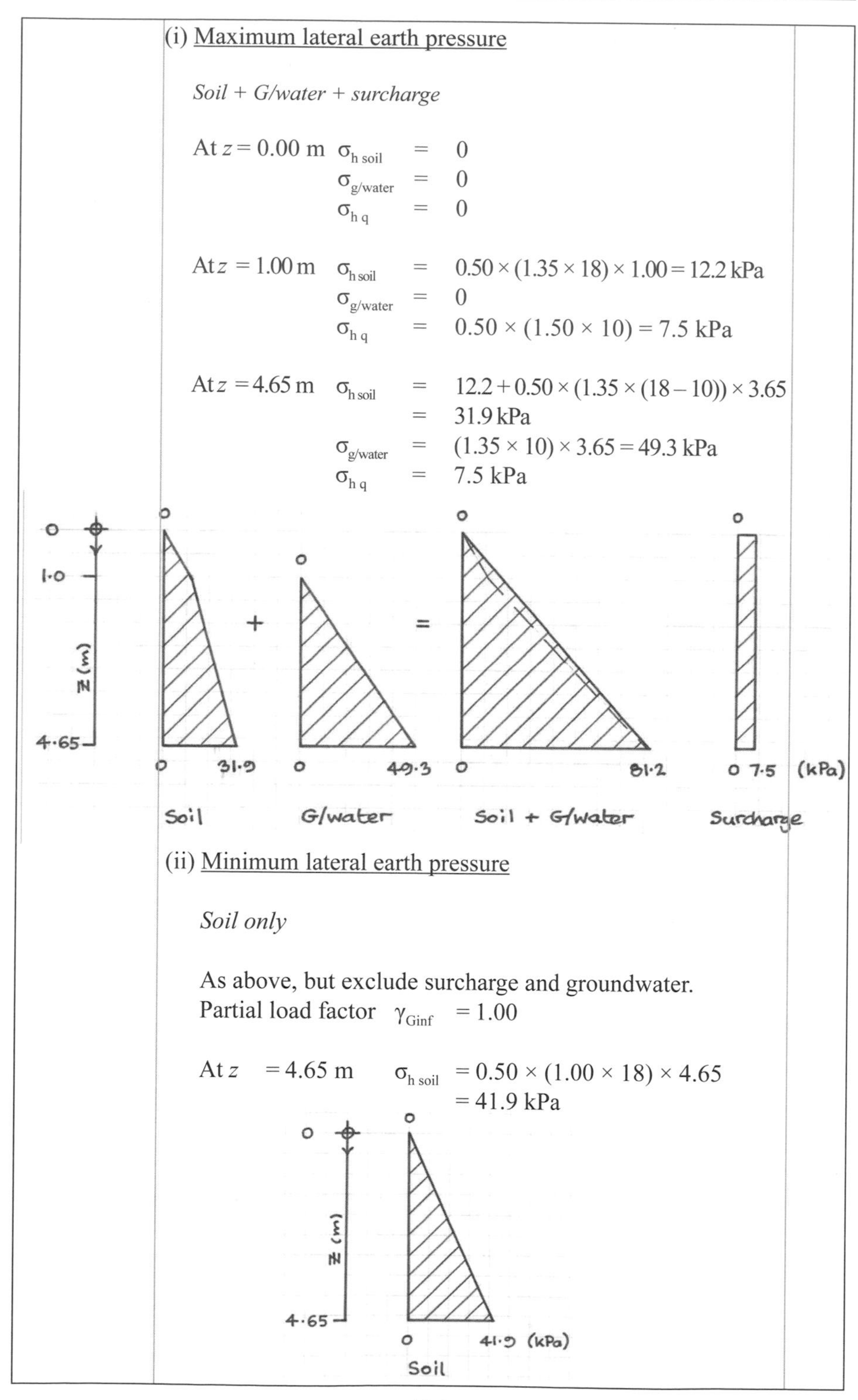

(i) Maximum lateral earth pressure

Soil + G/water + surcharge

At $z = 0.00$ m
$\sigma_{h\ soil} = 0$
$\sigma_{g/water} = 0$
$\sigma_{h\ q} = 0$

At $z = 1.00$ m
$\sigma_{h\ soil} = 0.50 \times (1.35 \times 18) \times 1.00 = 12.2$ kPa
$\sigma_{g/water} = 0$
$\sigma_{h\ q} = 0.50 \times (1.50 \times 10) = 7.5$ kPa

At $z = 4.65$ m
$\sigma_{h\ soil} = 12.2 + 0.50 \times (1.35 \times (18 - 10)) \times 3.65$
$= 31.9$ kPa
$\sigma_{g/water} = (1.35 \times 10) \times 3.65 = 49.3$ kPa
$\sigma_{h\ q} = 7.5$ kPa

(ii) Minimum lateral earth pressure

Soil only

As above, but exclude surcharge and groundwater.
Partial load factor $\gamma_{Ginf} = 1.00$

At $z = 4.65$ m
$\sigma_{h\ soil} = 0.50 \times (1.00 \times 18) \times 4.65$
$= 41.9$ kPa

Serviceability Limit State (SLS)

BS EN 1997-1 §2.4.8(2)

Partial factors on Actions

$\gamma_G = 1.00$

$\gamma_Q = 1.00$

BS EN 1997-1 §9.8.1(2)

Use characteristic values of soil parameters to derive SLS design earth pressures.

BS EN 1997 Eqn 9.1

Coefficient of Lateral Earth Pressure 'At Rest'

$K_0 = 1 - \sin \varphi'_k$

$= 1 - \sin 30° = 0.50$

Obtain lateral earth pressures using the same approach as for ULS.

(i) Maximum Lateral Earth Pressure

Soil + G/water + Surcharge

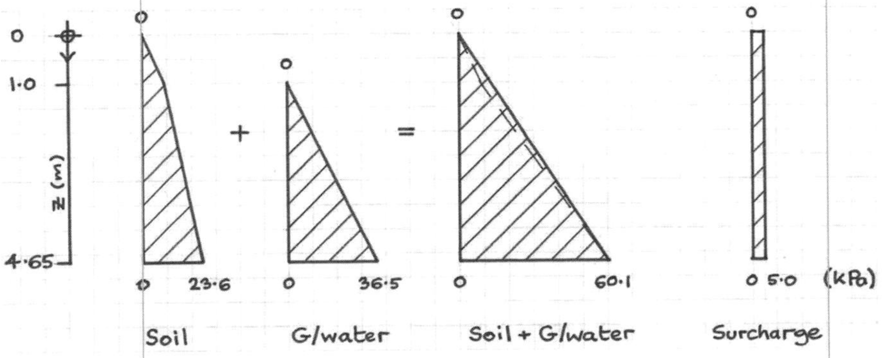

(ii) Minimum Lateral Earth Pressure

Soil only

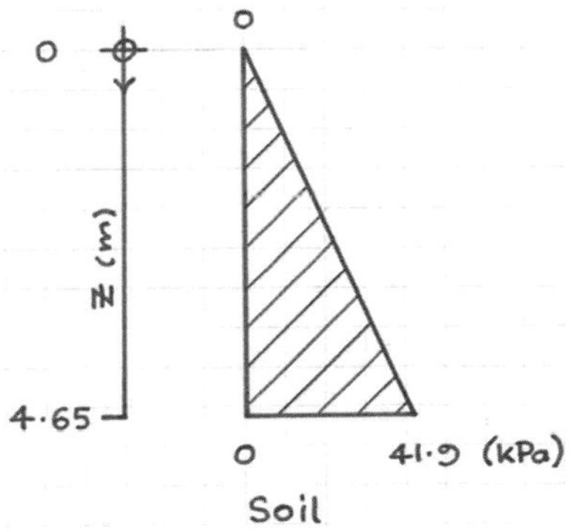

4. HYDROSTATIC WATER PRESSURES IN WET WELL

BS EN 1991-4

Determine hydrostatic pressures on walls of wet well for Ultimate Limit State (STR) and Serviceability Limit State (SLS) in accordance with BS EN 1991-4.

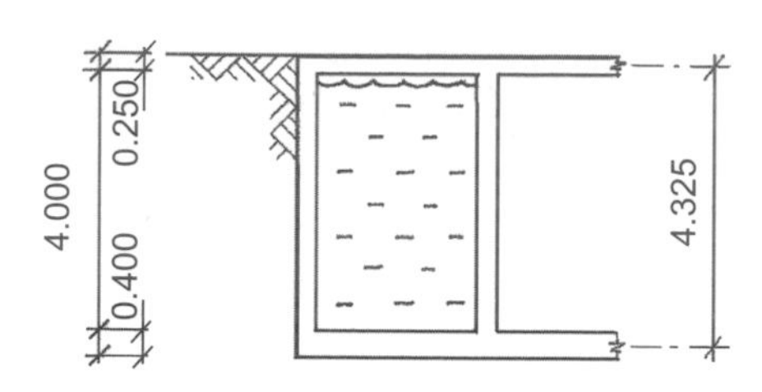

Basic data

Unit weight of water, γ_w = 10 kN/m³

Ultimate Limit State (STR)

BS EN 1991-4 §B.3

Partial factors on Actions

In operation, γ_F = 1.20

Under test, γ_F = 1.00

In operation At z = 4.325 m, σ_{water} = (1.20 × 10) × 4.325
= 51.9 kPa

Under test At z = 4.325 m, σ_{water} = (1.00 × 10) × 4.325
= 43.3 kPa

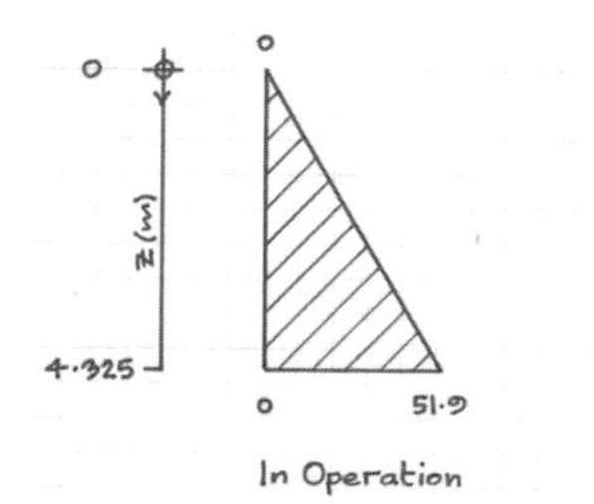

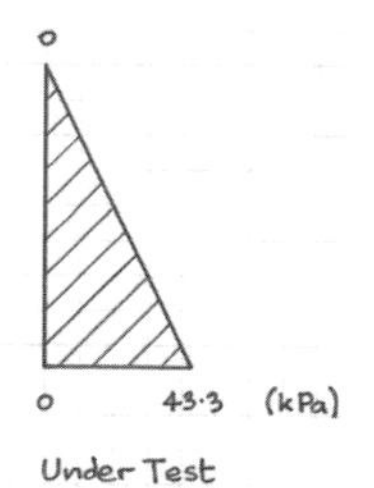

Serviceability Limit State

BS EN 1990 §A1.4.1(1)

Partial factors on Actions

In operation and under test, γ_F = 1.00

In operation and under test

At z = 4.325 m, σ_{water} = (1.00 × 10) × 4.325
= 43.3 kPa

5. BENDING MOMENTS IN WALLS

Calculate bending moments in walls for Ultimate Limit State (STR) and Serviceability Limit State (SLS) under soil, groundwater and hydrostatic pressure loadings using bending moment coefficients.

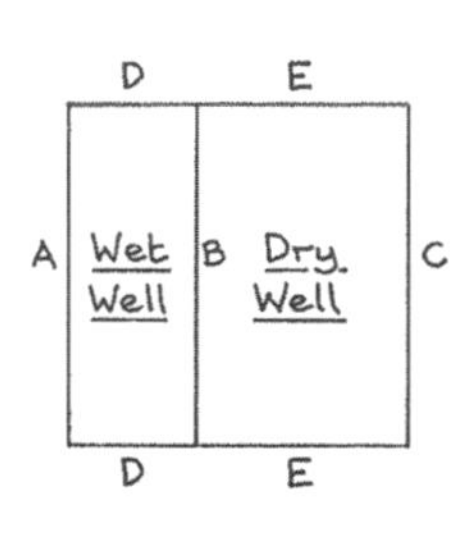

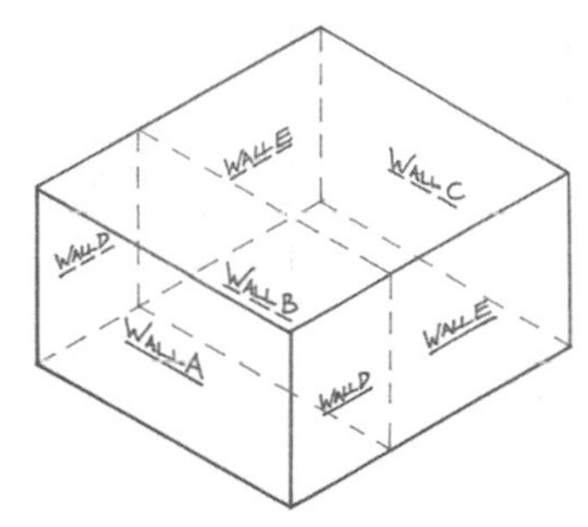

Analysis method

1. Consider walls as two-way spanning with continuous supports on all four sides.
2. Obtain bending moments (BMs) using coefficients for a rectangular two-way spanning slab with built-in/fixed edge supports and a triangular load distribution (elastic analysis).
   (Note: The coefficients used in the example are taken from Table 53 in the Reinforced Concrete Designer's Handbook (Tenth edition) by Reynolds & Steedman.)
3. Use lateral earth and hydrostatic pressure distributions as obtained in §§3 and 4 of these calculations.
4. Effective heights and widths of wall panels are measured from the centre lines of supporting elements (ref. Figure 6.1).
5. Bending moment sign convention:
   M_V vertical span direction
   M_H horizontal span direction
   $+M$ tension in unloaded face (span)
   $-M$ tension in loaded face (support)

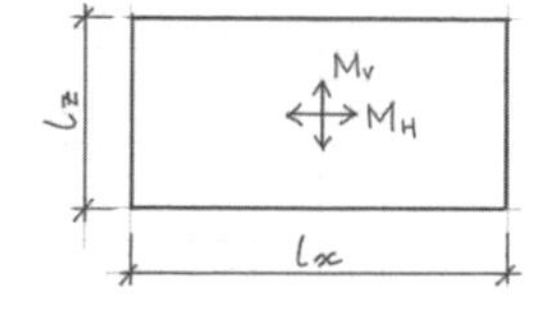

6. No elastic distribution of bending moments at wall intersections has been carried out. Conservative worst case load effects are thus considered in design.
7. Consider the following cases:
   Case 1–Maximum external soil and groundwater pressure.
   Case 2–Maximum internal water pressure, with 'wet well' full.
8. Consider the following critical construction stages (ref. Figure 6.2):
   Stage 5–Completed structure without backfill, during 'wet well' water test.
   Stage 6–Completed structure with backfill, in service.

WALL A

Geometry and bending moment coefficients

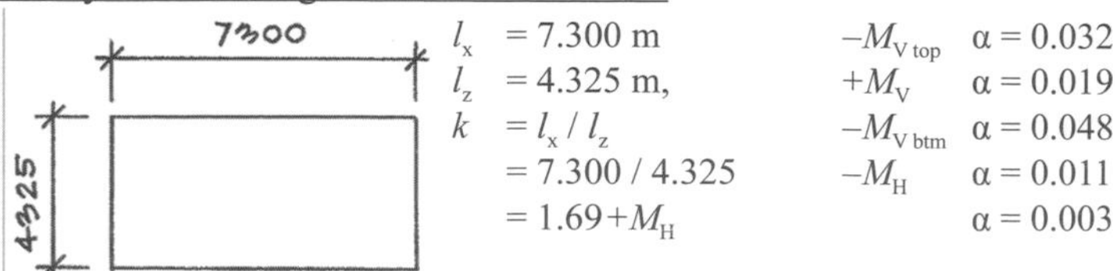

l_x = 7.300 m
l_z = 4.325 m,
$k = l_x / l_z$
= 7.300 / 4.325
= 1.69

$-M_{V\,top}$	$\alpha = 0.032$
$+M_V$	$\alpha = 0.019$
$-M_{V\,btm}$	$\alpha = 0.048$
$-M_H$	$\alpha = 0.011$
$+M_H$	$\alpha = 0.003$

Ultimate Limit State (STR)

Case 1–Max. External Soil & Water Pressure

Completed structure (Stage 6)

For ease of calculation, combine lateral soil and groundwater pressures into a single equivalent load. (Note. Apply this loading to all external walls.)

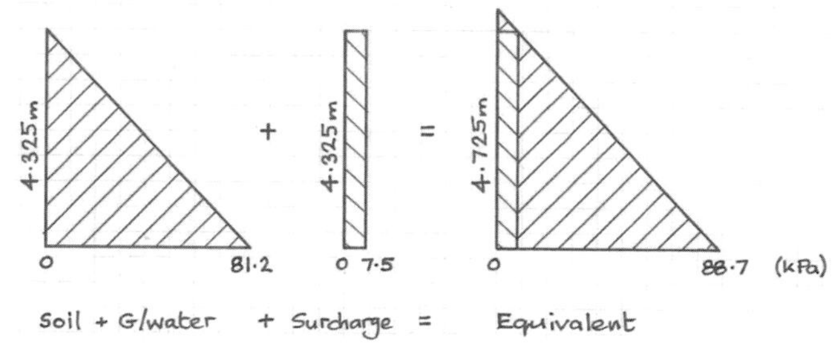

Bending moments Soil + Groundwater +Surcharge (σ_{max} = 88.7 kPa)

$-M_{V\,top} = 0.032 \times 88.7 \times 4.725 \times 4.325 = 58.0$ kNm/m
$+M_V = 0.019 \times 88.7 \times 4.725 \times 4.325 = 34.4$ kNm/m
$-M_{V\,btm} = 0.048 \times 88.7 \times 4.725 \times 4.325 = 87.0$ kNm/m
$-M_H = 0.011 \times 88.7 \times 7.300^2 = 52.0$ kNm/m
$+M_H = 0.003 \times 88.7 \times 7.300^2 = 14.2$ kNm/m

Case 2–Internal water pressure

(a) In operation Completed structure with backfill (Stage 6)

In operation, water pressure in the wet well will act against exterior lateral earth pressure to reduce the net loading on Wall A. By inspection this will not give a critical load combination.

(b) Under test Completed structure without backfill (Stage 5)

Bending moments Water (σ_{max} = 43.3 kPa)

$-M_{V\,top} = 0.032 \times 43.3 \times 4.325^2 = 25.9$ kNm/m
$+M_V = 0.019 \times 43.3 \times 4.325^2 = 15.4$ kNm/m
$-M_{V\,btm} = 0.048 \times 43.3 \times 4.325^2 = 38.9$ kNm/m
$-M_H = 0.011 \times 43.3 \times 7.300^2 = 25.4$ kNm/m
$+M_H = 0.003 \times 43.3 \times 7.300^2 = 6.9$ kNm/m

By comparison, bending moments from Case 1 are more onerous than from Case 2.

Serviceability Limit State (SLS)

Case 1–Max. External Soil & Water Pressure

Completed Structure (Stage 6)

For ease of calculation, combine lateral soil and groundwater pressures into a single equivalent load. (Note: apply this loading to all external walls.)

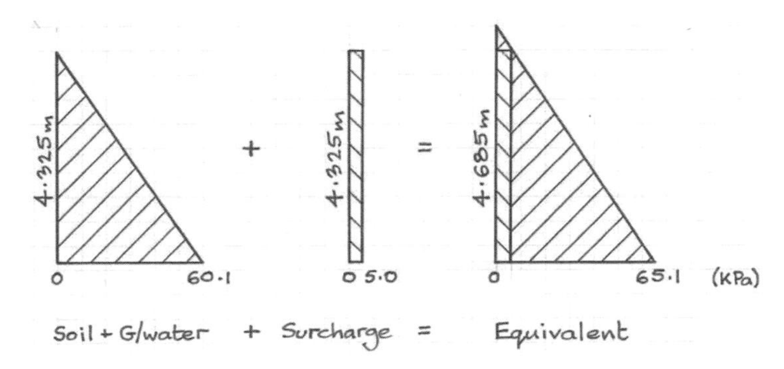

Bending moments Soil + Groundwater + Surcharge ($\sigma = 65.1$ kPa)

$-M_{V\,top} = 0.032 \times 65.1 \times 4.725 \times 4.325 = 42.6$ kNm/m
$+M_V = 0.019 \times 65.1 \times 4.725 \times 4.325 = 25.3$ kNm/m
$-M_{V\,btm} = 0.048 \times 65.1 \times 4.725 \times 4.325 = 63.9$ kNm/m
$-M_H = 0.011 \times 65.1 \times 7.300^2 = 38.2$ kNm/m
$+M_H = 0.003 \times 65.1 \times 7.300^2 = 10.4$ kNm/m

Case 2–Internal water pressure

(a) In operation Completed structure with backfill (Stage 6)
As for ULS, in operation water pressure in the wet well will act against exterior lateral earth pressure to reduce the net loading on Wall A. By inspection this will not give a critical load combination.

(b) Under test Completed structure without backfill (Stage 5)

Bending moments Water ($\sigma_{max} = 43.3$ kPa)

$-M_{V\,top} = 0.032 \times 43.3 \times 4.325^2 = 25.9$ kNm/m
$+M_V = 0.019 \times 43.3 \times 4.325^2 = 15.4$ kNm/m
$-M_{V\,btm} = 0.048 \times 43.3 \times 4.325^2 = 38.9$ kNm/m
$-M_H = 0.011 \times 43.3 \times 7.300^2 = 25.4$ kNm/m
$+M_H = 0.003 \times 43.3 \times 7.300^2 = 6.9$ kNm/m

By comparison, bending moments from Case 1 are more onerous than from Case 2.

WALL B

Geometry and bending moment coefficients

Same as WALL A.

Ultimate Limit State (STR)

Case 1–Max. External Soil & Water Pressure–Not applicable.

Case 2–Internal Water Pressure

(a) In Operation   Completed structure with backfill (Stage 6)

Bending Moments   Water (σ_{max} = 51.9 kPa)

$-M_{V\,top}$ = 0.032 × 51.9 × 4.325² = 31.1 kNm/m
$+M_V$ = 0.019 × 51.9 × 4.325² = 18.4 kNm/m
$-M_{V\,btm}$ = 0.048 × 51.9 × 4.325² = 46.6 kNm/m
$-M_H$ = 0.011 × 51.9 × 7.300² = 30.4 kNm/m
$+M_H$ = 0.003 × 51.9 × 7.300² = 8.3 kNm/m

(b) Under Test   Completed structure without backfill (Stage 5)

Bending Moments   Water (σ_{max} = 43.3 kPa)

By inspection, bending moments for 'In Operation' case will be more onerous.

Serviceability Limit State (SLS)

Case 1–Max. External Soil & Water Pressure–Not applicable.

Case 2–Internal Water Pressure

(a) In Operation   Completed structure with backfill (Stage 6)

Bending Moments   Water (σ_{max} = 43.3 kPa)

$-M_{V\,top}$ = 0.032 × 43.3 × 4.325² = 25.9 kNm/m
$+M_V$ = 0.019 × 43.3 × 4.325² = 15.4 kNm/m
$-M_{V\,btm}$ = 0.048 × 43.3 × 4.325² = 38.9 kNm/m
$-M_H$ = 0.011 × 43.3 × 7.300² = 25.4 kNm/m
$+M_H$ = 0.003 × 43.3 × 7.300² = 6.9 kNm/m

(b) Under Test   Completed structure without backfill (Stage 5)

Bending Moments   Water (σ_{max} = 43.3 kPa)

Bending moments are identical to 'In Operation' case.

WALL C

Bending moments for Wall C are the same calculated as for Wall A at ULS and SLS, with the exception that Case 2 (Internal Water Pressure) does not apply.

WALL D

Geometry and bending moment coefficients

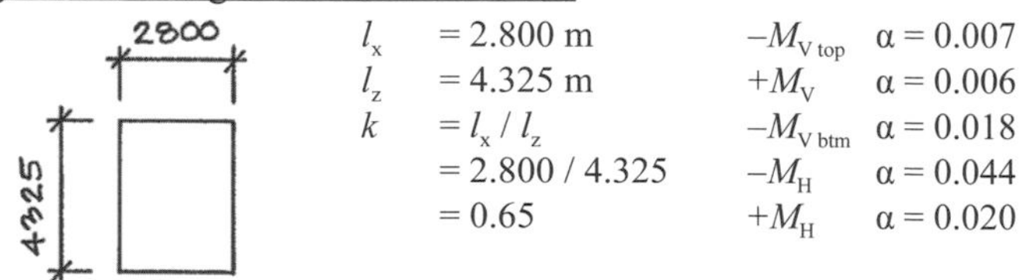

l_x = 2.800 m
l_z = 4.325 m
$k = l_x / l_z$
= 2.800 / 4.325
= 0.65

$-M_{V\,top}$ α = 0.007
$+M_V$ α = 0.006
$-M_{V\,btm}$ α = 0.018
$-M_H$ α = 0.044
$+M_H$ α = 0.020

Ultimate Limit State (STR)

Case 1–Max. External Soil & Water Pressure

Completed structure (Stage 6)

Bending moments Soil + Groundwater + Surcharge (σ_{max} = 88.7 kPa)

$-M_{V\ top}$ = 0.007 × 88.7 × 4.725 × 4.325 = 12.7 kNm/m
$+M_V$ = 0.006 × 88.7 × 4.725 × 4.325 = 10.9 kNm/m
$-M_{V\ btm}$ = 0.018 × 88.7 × 4.725 × 4.325 = 32.6 kNm/m
$-M_H$ = 0.044 × 88.7 × 2.800² = 35.6 kNm/m
$+M_H$ = 0.020 × 88.7 × 2.800² = 13.9 kNm/m

Case 2–Internal water pressure

(a) In operation Completed structure with backfill (Stage 6)

Water pressure in the wet well will act against exterior lateral earth pressure to reduce the net loading on Wall A. By inspection this will not give a critical load combination.

(b) Under test Complete structure without backfill (Stage 5)

Bending moments Water (σ_{max} = 43.3 kPa)

$-M_{V\ top}$ = 0.007 × 43.3 × 4.325² = 5.7 kNm/m
$+M_V$ = 0.006 × 43.3 × 4.325² = 4.9 kNm/m
$-M_{V\ btm}$ = 0.018 × 43.3 × 4.325² = 14.6 kNm/m
$-M_H$ = 0.044 × 43.3 × 2.800² = 14.9 kNm/m
$+M_H$ = 0.020 × 43.3 × 2.800² = 6.8 kNm/m

Serviceability Limit State (SLS)

Case 1–Max. External Soil & Water Pressure

Completed structure (Stage 6)

Bending moments Soil + Groundwater + Surcharge (σ = 65.1 kPa)

$-M_{V\ top}$ = 0.007 × 65.1 × 4.725 × 4.325 = 9.3 kNm/m
$+M_V$ = 0.006 × 65.1 × 4.725 × 4.325 = 8.0 kNm/m
$-M_{V\ btm}$ = 0.018 × 65.1 × 4.725 × 4.325 = 23.9 kNm/m
$-M_H$ = 0.044 × 65.1 × 2.800² = 22.5 kNm/m
$+M_H$ = 0.020 × 65.1 × 2.800² = 10.2 kNm/m

Case 2–Internal water pressure

(a) In operation Not critical (as for ULS).

(b) Under test Complete structure without backfill (Stage 5)

Bending moments Water (σ_{max} = 43.3 kPa)

$-M_{V\ top}$ = 0.007 × 43.3 × 4.325² = 5.7 kNm/m
$+M_V$ = 0.006 × 43.3 × 4.325² = 4.9 kNm/m
$-M_{V\ btm}$ = 0.018 × 43.3 × 4.325² = 14.6 kNm/m
$-M_H$ = 0.044 × 43.3 × 2.800² = 14.9 kNm/m
$+M_H$ = 0.020 × 43.3 × 2.800² = 6.8 kNm/m

WALL E

Geometry and bending moment coefficients

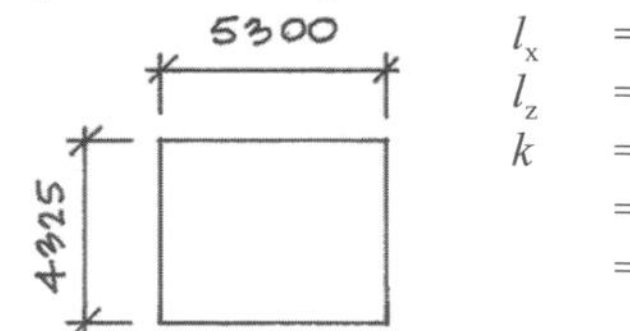

l_x = 2.800 m
l_z = 4.325 m
k = l_x / l_z
= 2.800 / 4.325
= 0.65

$-M_{V\,top}$ α = 0.007
$+M_V$ α = 0.006
$-M_{V\,btm}$ α = 0.018
$-M_H$ α = 0.044
$+M_H$ α = 0.020

Ultimate Limit State (STR)

Case 1–Max. External Soil & Water Pressure

Completed structure (Stage 6)

Bending moments Soil + Groundwater + Surcharge (σ_{max} = 88.7 kPa)

$-M_{V\,top}$ = 0.015 × 88.7 × 4.325² = 27.1 kNm/m
$+M_V$ = 0.024 × 88.7 × 4.325² = 43.5 kNm/m
$-M_{V\,btm}$ = 0.041 × 88.7 × 4.325² = 74.3 kNm/m
$-M_H$ = 0.019 × 81.2 × 5.300² = 47.3 kNm/m
$+M_H$ = 0.007 × 81.2 × 5.300² = 17.4 kNm/m

Case 2–Internal water pressure–Not applicable.

Serviceability Limit State (SLS)

Case 1–External soil and water pressure

Completed structure (Stage 6)

Bending moments Soil + Groundwater + Surcharge (σ = 65.1 kPa)

$-M_{V\,top}$ = 0.015 × 65.1 × 4.725 × 4.325 = 20.9 kNm/m
$+M_V$ = 0.024 × 65.1 × 4.725 × 4.325 = 30.1 kNm/m
$-M_{V\,btm}$ = 0.041 × 65.1 × 4.725 × 4.325 = 50.1 kNm/m
$-M_H$ = 0.019 × 65.1 × 5.300² = 35.1 kNm/m
$+M_H$ = 0.007 × 65.1 × 5.300² = 14.0 kNm/m

Case 2–Internal water pressure–Not applicable.

SUMMARY–ULS and SLS bending moments in walls (worst cases)

Effect	Bending Moment (kNm/m)										Description
	Wall A		Wall B		Wall C		Wall D		Wall E		
	ULS	SLS	ULS	SLS	ULS	SLS	ULS	SLS	ULS	SLS	
$-M_{V\,top}$	58.0	42.6	31.1	25.9			12.7	9.3	27.1	20.9	Roof /wall junction
$+M_V$	34.4	25.3	18.4	15.4			10.9	8.0	43.5	30.1	Wall panel (vertical)
$-M_{V\,btm}$	87.0	63.9	46.6	38.9	As Wall A		32.6	23.9	74.3	50.1	Base/wall junction
$-M_H$	52.0	38.2	30.4	25.4			35.6	22.5	47.3	35.1	Wall/wall junction
$+M_H$	14.2	10.4	8.3	6.9			13.9	10.2	17.4	14.0	Wall panel (horizontal)

6. DESIGN ACTIONS ON BASE SLAB

Determine the bearing pressure on base slab due to gravity loads (uniform distribution assumed). The exact distribution of bearing pressure on the base slab will depend on soil-structure interaction. However, this approach gives moderately conservative bending moments in the base slab at mid-span and is reasonable for preliminary design.

Geometry

Area of base slab = 8.4 × 7.6 = 63.8 m^2

Dead load

Sheet PH/3

Weight of reinforced concrete structure = 2171 kN

Weight of mass concrete infill to wet well

= 24.0 × [0.5 × (7.0 × 2.5 × 1.5)] = 315 kN

Total = 2486 kN

Bearing pressure from Dead Loads

= 2486 / 63.8 = 39.0 kPa

Imposed load

Weight of water in wet well

= (7.0 × 2.5) × (4.0 – 1.5/2) × 10.0 = 569 kN

Weight of pumps, pipework, controls, etc. (approx.) = 100 kN

Live load on control room floor (assume 3.0 kPa UDL)

= 8.4 × 7.6 × 3.0 = 192 kN

Total = 861 kN

Bearing pressure from Imposed Loads

= 861 / 63.8 = 13.5 kPa

Ultimate Limit State (STR)

Partial Factors on Actions

$\gamma_{G,sup}$ = 1.35, γ_Q = 1.50

Worst Case Bearing Pressure

w = (1.35 × 39.0) + (1.50 × 13.5) = 72.9 kPa

Serviceability Limit State (SLS)

Partial Factors on Actions

γ_{Gsup} = 1.00, γ_Q = 1.00

Worst Case Bearing Pressure

V = (1.00 × 39.0) + (1.00 × 13.5) = 52.5 kPa

7. BENDING MOMENTS IN BASE SLAB

Calculate bending moments in base slab for Ultimate Limit State (STR) and Serviceability Limit State (SLS) using bending moment coefficients and fixing moments from walls.

Analysis method

1. Consider the bearing pressure on base slab due to gravity loads obtained in §6.
2. Apply the bearing pressure to the base slab as a UDL.
3. Obtain the 'net' bending moments in the slab by considering the 'free' bending moments for a slab with simple supports and adding the effects of the minimum 'fixing' BMs provided by the perimeter walls of the continuous structure.
4. Determine 'free' bending moments for unit widths of slab in short-span and long-span directions assuming edges of slab to be simply supported. Obtain BMs using coefficients for a rectangular two-way spanning slab with simply supported edges and uniformly distributed load (elastic analysis).
   (Note: The coefficients used in the example are taken from Table 50 in the Reinforced Concrete Designer's Handbook (Tenth edition) by Reynolds & Steedman.)
5. Take account of continuity between slab and perimeter walls by adding the effects of coexistent 'fixing' moments due to lateral earth pressures acting on the walls.
6. By inspection, bending moments in the base slab panel to the dry well will be more onerous than those in the panel to the wet well. Thus only bending moments in the dry well base slab panel will be calculated.

DRY WELL BASE SLAB

Geometry and bending moment coefficients

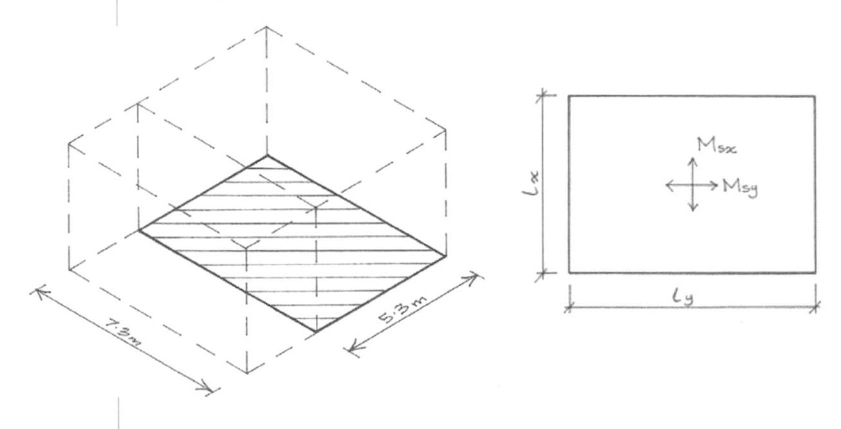

Bending moment coefficients for UDL on base slab to dry well slab (simply supported)

l_x = 5.300 m

l_y = 7.300 m

k = l_y / l_x = 7.300 / 5.300 = 1.38

Bending moment coefficients are

α_{sx} = 0.783, α_{sy} = 0.217

Ultimate Limit State (STR)

Bending moments (Bearing pressure = 72.9 kPa)

Obtain the 'net' BMs in the slab by considering the 'free' BMs for a slab with simple supports and adding the effects of the minimum 'fixing' BMs provided by lateral earth pressure acting on the perimeter walls C and E of the continuous structure.

'Free' BMs for simply supported slab

M_{sx} = $0.783 \times 72.9 \times 5.300^2 / 8$ = 200.4 kNm/m

M_{sy} = $0.217 \times 72.9 \times 7.300^2 / 8$ = 105.4 kNm/m

'Fixing' BMs

Sheet PH/7

The minimum lateral earth pressure at z = 4.65 m = 41.9 kPa

Use BM coefficients for Walls C and E previously obtained.

Fixing BM to M_{sx} is provided by Wall C.

Sheet PH/11

BM coefficient, α = 0.048

Min. 'Fixing' BM = $0.048 \times 41.9 \times 4.325^2$ = 37.6 kNm/m

Fixing BM to M_{sy} is provided by Wall E.

Sheet PH/15

BM coefficient, α = 0.041

Min. 'Fixing' BM = $0.041 \times 41.9 \times 4.325^2$ = 32.1 kNm/m

'Net' BMs in Slab

At mid-span of short span,

'Net' M_{sx} = 'Free' M_{sx} + 'Fixing' M_{sx} = 200.4 + (−37.6/2) = 181.6 kNm/m

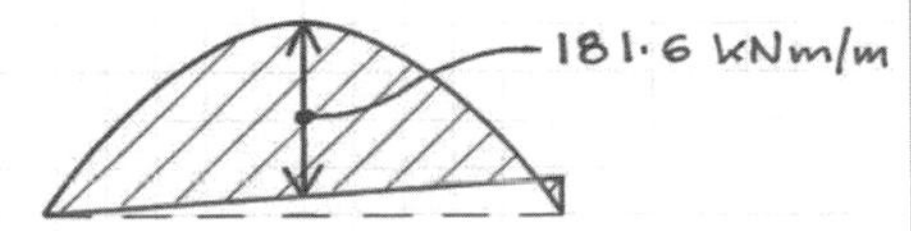

At mid-span of long span,

'Net' M_{sy} = 'Free' M_{sy} + 'Fixing' M_{sy} = 105.4 + −32.1 = 73.3 kNm/m

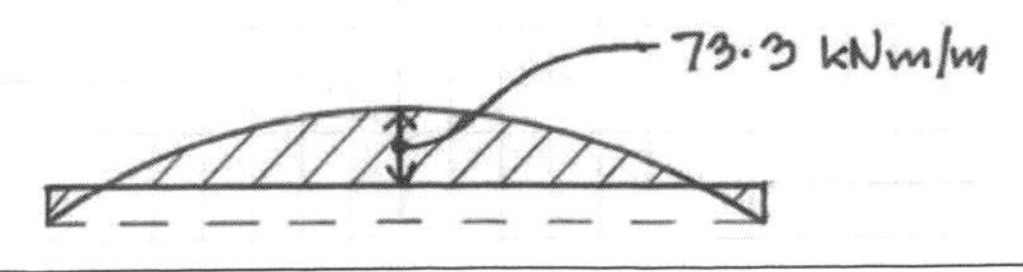

Serviceability Limit State (SLS)

Bending moments (Bearing pressure = 52.5 kPa)

'Free' BMs for simply supported slab

M_{sx} = $0.783 \times 52.5 \times 5.300^2 / 8$ = 144.3 kNm/m

M_{sy} = $0.217 \times 52.5 \times 7.300^2 / 8$ = 75.9 kNm/m

'Fixing' BMs

Sheet PH/8

The minimum lateral earth pressure at z = 4.65 m = 41.9 kPa

Fixing BM to M_{sx} is provided by Wall C.

Min. 'Fixing' BM = $0.048 \times 41.9 \times 4.325^2$ = 37.6 kNm/m

Fixing BM to M_{sy} is provided by Wall E.

Min. 'Fixing' BM = $0.041 \times 41.9 \times 4.325^2$ = 32.1 kNm/m

Net BMs in Slab

At mid-span of short span,

'Net' M_{sx} = 'Free' M_{sx} + 'Fixing' M_{sx} = 144.3 + (−37.6/2) = 125.5 kNm/m

BMD

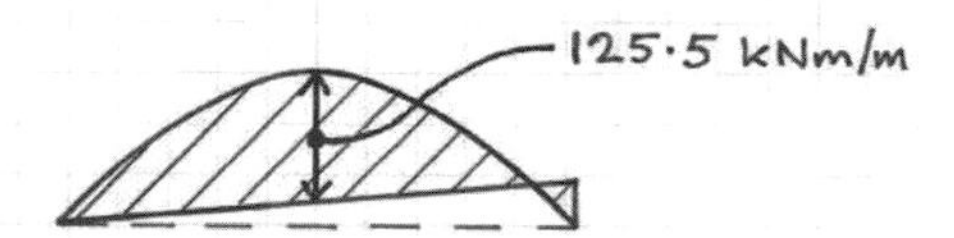

At mid-span of long span,

'Net' M_{sy} = 'Free' M_{sy} + 'Fixing' M_{sy} = 75.9 + −32.1 = 43.8 kNm/m

BMD

SUMMARY–ULS and SLS bending moments in dry well base slab

Effect	Bending Moment (kNm/m)			
	Midspan		Slab/Wall Junction*	
	ULS	SLS	ULS	SLS
M_{sx} (in short span direction)	181.6	125.5	(−) 87.0	(−) 63.9
M_{sy} (in long span direction)	73.3	43.8	(−) 46.6	(−) 38.9

* Design slab at junctions with walls for maximum wall BMs as these are larger than the slab moments (see summary table on sheet PH/15).

8. SLS & ULS DESIGN CHECKS FOR SAMPLE WALL AND BASE SLAB PANELS

BS EN 1992-1-1
BS EN 1992-3

Design calculations for verifications at Serviceability Limit State (SLS) and at Ultimate Limit State (ULS-STR) are given for the following elements of the buried box structure:

- vertical reinforcement to Wall A at junction with Base Slab
- horizontal reinforcement to Wall A at junction with Wall D
- main reinforcement to Dry Well panel of Base Slab

Material properties

Concrete

BS EN 1992-1-1 Table 3.1

Strength Class C28/35 f_{ck} = 28 MPa (cylinder strength)

f_{ctm} = 2.8 MPa

Modulus (short term) E_{cm} = 32 GPa

BS EN 1992-1-1 §3.1.4

Creep coefficient.

Age of concrete at first loading, t_0 = 50 days

Long-term relative humidity, RH = 80%

For 300 mm thick wall (worst case),

$h_0 = 2 \times (1\,000 \times 300) / (2 \times 1\,000)$ = 300 mm

BS EN 1992-1-1 Fig. 3.1(b)

Hence creep coefficient, $\varphi(\infty, t_0)$ = 1.6

Modulus (long term),

$E_{c,eff} = E_c/(1 + \varphi) = 1.05 \times E_{cm}/(1 + \varphi) = 1.05 \times 32 / 2.6$

$= 12.9$ GPa

Steel reinforcement

BS EN 1992-1-1 §3.2.7(4)

Grade B500B to BS 8666 f_{yk} = 500 MPa

E_s = 200 GPa

Minimum reinforcement areas

Walls

BS EN 1992-1-1 §9.6.2 (1)

$A_{s,vmin} \geq 0.002 \times A_c$

$\geq 0.002 \times 300 \times 1\,000$ = 600 mm²/m

Base slabs

BS EN 1992-1-1 §9.3.1.1 (1)

$A_{s,vmin} \geq 0.26 \times (f_{ctm}/f_{yk}) \times b_t \times d$

For 400 thick base slab,

$d_{max} = 400 - 40 - (20/2)$ = 350 mm

$A_{s,vmin} \geq 0.26 \times (2.8/5\,000) \times 1\,000 \times 340$

$= 0.00146 \times 1\,000 \times 350$ = 511 mm²/m

WALL A–Vertical reinforcement at junction with base slab

The following principal design verifications will be made:

Serviceability Limit State (SLS)

- Flexural crack width
- Stress limitations
- Deflection

Ultimate Limit State (ULS)

- Bending
- Shear

Serviceability limit state (SLS)

Flexural crack width

BS EN 1992-3 §7.3.1

Adopt 'Tightness Class 1'

Determine limiting crack width, w_{k1}

$h_D = 3.650 - 0.400 = 3.250$ m

$h = 0.300$ m

$h_D/h = 3.250 / 0.300 = 10.8$

Obtain w_{k1} by interpolation between values given in Note to §7.3.1(111)

$w_{k1} = 0.2 - (10.8 - 5) \times (0.2 - 0.05)/(35 - 5)$

$= 0.17$ mm

Verify crack width for worst-case bending moment at base of Wall A (Case 1–Maximum external soil and water pressure).

Sheet PH/15

$M_{Ed} = M_{V\,btm} = 63.9$ kNm/m

BS EN 1992-1-1 §7.3.4

Calculate crack width, w_k

Assume main reinforcement B20 bars @150 mm spacing

$A_s = 2095$ mm²/m

$b = 1\,000$ mm

$d = 300 - 40 - 16 - (20/2) = 234$ mm

To determine stress in reinforcement, first need to calculate neutral axis depth, x. From analysis of cracked cross section with a triangular concrete stress block,

$x = \{-\alpha_e.A_s + \sqrt{[(\alpha_e.A_s)^2 + 2.b.(\alpha_e.A_s).d]}\}/b$

$\alpha_e = E_s/E_{c,eff} = 200 / 12.9 = 15.5$

$\alpha_e.A_s = 15.5 \times 2095 = 32473$ mm²

$x = \{-32473 + \sqrt{[(32473)^2 + 2.1000.(32473).234]}\}/1\,000 = 95$ mm

$z = d - x/3 = 234 - 95/3 = 202$ mm

Determine $\varepsilon_{sm} - \varepsilon_{cm}$

$F_s = M_{Ed}/z = 63.9 / 0.202 = 316$ kN

$\sigma_s = 316 \times 10^3 / 2095 = 151$ MPa

$k_t = 0.4$ (long-term loading)

$f_{ct,eff} = f_{ctm} = 2.8$ MPa

$h_{c,eff} = (300 - 95) / 3 = 68$ mm

$\rho_{p,eff} = 2095 / (68 \times 1\,000) = 0.031$

BS EN 1992-1-1 Eqn (7.9)

$\varepsilon_{sm} - \varepsilon_{cm} = [151 - 0.4.(2.8/0.031).(1 + (32).(0.031)] / 200\,000 = 0.00040$

(Minimum value = $0.6 \times 149 / 200\,000 = 0.00045$)

Determine $s_{r,max}$

$\varphi = 20$ mm

$c = 40 + 16 = 56$ mm

Check maximum limit of bar spacing

$= 5 \times (56 + 20/2) = 330$ mm > actual bar spacing

$k_1 = 0.8$

$k_2 = 0.5$

$k_3 = 3.4$

$k_4 = 0.425$

BS EN 1992-1-1 Eqn (7.11)

$s_{r,max} = 3.4(56) + 0.8.(0.5).0.425.(20)/0.031 = 300$ mm

Determine w_k

BS EN 1992-1-1 Eqn (7.8)

$w_k = s_{r,max} \cdot (\varepsilon_{sm} - \varepsilon_{cm}) = 300 \times 0.00045 = 0.14$ mm

$w_{k1} = 0.17$ mm

$\boldsymbol{w_k < w_{k1}}$

Flexural crack width satisfactory for B20 bars @ 150 mm spacing

(Note: Cracking due to thermal and shrinkage effects must also be verified using the methodology in §8. This may affect the reinforcement required.)

Stress limitations

BS EN 1992-1-1 §7.2

Verify concrete and steel stresses for worst-case bending moment at base of Wall A (Case 1–Maximum external soil and water pressure).

Concrete.

Limiting compressive stress

$= k_2 \times f_{ck} = 0.45 \times 28 = 12.6$ MPa

$F_c = F_s = 316$ kN

$\sigma_c = 316 \times 10^3 / (0.5 \times 95 \times 1\,000) = 7.1$ MPa

< limiting stress

Reinforcing steel.

Limiting tensile stress

$= k_3 \times f_{yk} = 0.8 \times 500 = 400$ MPa

As previously calculated, $\sigma_s = 151$ MPa

$<$ limiting stress

Limiting stresses satisfactory

Deflection

BS EN 1992-1-1 §7.4.2 (2)

Verify acceptability of horizontal deflection of Wall A by consideration of the limiting span/depth ratio.

Assume that at mid-height, the main flexural reinforcement has been curtailed to B16 bars @ 150 mm spacing.

$A_s = 1\,341$ mm²/m

$b = 1\,000$ mm

$d = 300 - 40 - 16 - (16/2) = 236$ mm

$\rho_0 = \sqrt{28} \times 10^{-3} = 0.0054$

$\rho = 1\,341/(1\,000 \times 236) = 0.0057$

$\rho > \rho_0$

so use Eq. (7.16.b) to determine limiting l/d ratio.

$K = 1.3$

$\rho' = 0$

BS EN 1992-1-1 Eqn (7.16.b)

Limiting l/d ratio

$= 1.3 \times [11 + 1.5 \times \sqrt{28} \times (0.0054/(0.0057 - 0)) + 0]$

$= 24.0$

$l/d = 4.325 / 0.236 = 18.3 <$ limiting ratio

Deflection satisfactory

Ultimate Limit State (STR)

Bending

Verify main reinforcement for worst-case bending moment at base of Wall A. (Case 1–Maximum external soil and water pressure).

Sheet PH/15

$M_{Ed} = M_{V\ btm} = 87.0$ kNm/m

Verify for B20 bars @ 150 mm pitch.

$A_s = 2095$ mm²/m

$b = 1\ 000$ mm

$d = 300 - 40 - 16 - (20/2) = 234$ mm

$f_{ck} = 28$ MPa

$\gamma_c = 1.50$

$f_{yk} = 500$ MPa

$\gamma_s = 1.15$

Use rectangular stress block.

$k = M_{Ed} / (b \times d^2 \times f_{ck})$

$= 87.0 \times 10^6 / (1\ 000 \times 234^2 \times 28) = 0.057$

Lever arm factor

$= 0.5 + \sqrt{(0.25 - k/1.14)}$

$= 0.5 + \sqrt{(0.25 - 0.057/1.14)} = 0.947$

(Limiting value $= 0.950$)

$z = 0.947 \times 234 = 221$ mm

$A_{s\ reqd} = M_{Ed} / (f_{yd} \times k)$

$= M_{Ed} / ((1/\gamma_s) \times f_{yk} \times z)$

$= 87.0 \times 10^6 / (0.87 \times 500 \times 221)$

$= 905$ mm²/m $> A_{s\ min}$

$A_{s\ prov} = 2095$ mm²/m

$\mathbf{A_{s\ prov} > A_{s\ reqd}}$

Bending satisfactory for B20 bars @ 150 mm spacing

Shear

Verify shear at the base of Wall A for worst-case horizontal loading (Case 1–Maximum external soil and water pressure).

As a conservative approximation, consider a 1.0 m wide strip of wall acting as a propped cantilever with triangular lateral earth pressure loading.

At root of propped cantilever, reaction force = 0.8 × total load on span

Sheet PH/11

Lateral earth pressure at base of wall = 88.7 kPa

Load = 88.7 × 0.5 × [(4.65 − 0.4) × 1.0] = 189 kN/m

V_{Ed} = 0.8 × 189 = 151 kN/m

Verify shear capacity for main tension reinforcement <u>B20 bars @ 150 mm spacing.</u>

BS EN 1992-1-1 §6.2.2

Determine $V_{Rd,c}$

$C_{Rd,c} = 0.18 / \gamma_c$

$= 0.18 / 1.50 = 0.12$

$k = 1 + \sqrt{(200 / 234)} = 1.92$

$\rho_l = 2\,095 / (1\,000 \times 234) = 0.009$

$<$ limiting value (0.02)

$V_{Rd,c} = [0.12 \times 1.92 \times (100 \times 0.009 \times 28)^{1/3}] \times 1\,000 \times 234 \times 10^{-3}$

$= 158$ kN/m

$V_{Ed} = 151$ kN/m

$\boldsymbol{V_{Rd,c} > V_{Ed}}$

<u>No design shear reinforcement required</u>

BS EN 1992-1-1 §6.2.3 (7)

Verify additional tensile force in longitudinal reinforcement due to shear.

$\Delta F_{td} \leq 0.5 \times 151 \times (2.5 - 0) = 189$ kN

$A_{s\ reqd} = 189 \times 10^3 / (0.87 \times 500) = 434$ mm²/m

Total area of longitudinal reinforcement required (bending + shear)

$A_{s\ reqd} = 905 + 434 = 1339$ mm²/m

$A_{s\ prov} = 2095$ mm²/m

$\boldsymbol{A_{s\ prov} > A_{s\ reqd}}$

<u>B20 @ 150 mm spacing are satisfactory</u>

<u>Summary</u>

<u>Main reinforcement of B20 bars @ 150 mm spacing is verified for SLS and ULS.</u>

WALL A / WALL D–Horizontal reinforcement at vertical junction between walls

Serviceability Limit State (SLS)

Sheet PH/14

$$M_{\text{Ed}} = M_{\text{H}} = 38.2 \text{ kNm/m}$$

Ultimate Limit State (STR)

Sheet PH/14

$$M_{\text{Ed}} = M_{\text{H}} = 52.0 \text{ kNm/m}$$

Use the same procedures as for Wall A to make verifications for SLS and ULS.

Shear

Flexural shear at the wall/wall junction should be checked, but calculation of the shear force distribution is beyond the scope of this example.

Direct tension

During water testing of the wet well (Construction Stage 5), the junction between Wall A and Wall D will be subject to direct tension because the walls will not be supported externally by the backfill.

This tensile force should be taken into account in the design of the horizontal reinforcement in the junction of these walls for the appropriate loading case (Case 2–Internal Water Pressure (Wet Well).

On the assumption that these checks will prove satisfactory when carried out, provide B16 bars @ 150 mm spacing.

BASE SLAB–Panel to dry well/pump room

Bending in short span direction at midspan

Serviceability Limit State (SLS)

Flexural crack width

BS EN 1992-3 §7.3.1

Adopt 'Tightness Class 1'

Determine limiting crack width w_{k1}.

h_D = 3.650 m

h = 0.400 m

h_D/h = 3.650 / 0.400 = 9.1

Obtain w_{k1} by interpolation between values given in Note to §7.3.1(111)

w_{k1} = 0.2 – (9.1 – 5) × (0.2 – 0.05)/(35 – 5)

= 0.18 mm

Verify crack width for worst-case bending moment in short span direction at mid-span.

Sheet PH/19

$M_{Ed} = M_{sx}$ = 125.5 kNm/m (hogging)

BS EN 1992-1-1 §7.3.4

Calculate crack width, w_k

Assume main reinforcement B25 bars @ 150 mm spacing

A_s = 3275 mm²/m

b = 1 000 mm

d = 400 – 40 – (25/2) = 347 mm

To determine stress in reinforcement, first need to calculate neutral axis depth, x.

From analysis of cracked cross section with a triangular concrete stress block,

$x = \{-\alpha_e.A_s + \sqrt{[(\alpha_e.A_s)^2 + 2.b.(\alpha_e.A_s).d]}\}/b$

$\alpha_e = E_s/E_{c,eff}$ = 200 / 12.9 = 15.5

$\alpha_e.A_s$ = 15.5 × 3275 = 50763 mm²

x = {–50763 + √[(50763)² + 2.1000.(50763).347]}/1 000 = 144 mm

$z = d - x/3$ = 347 – 144/3 = 299 mm

Determine $\varepsilon_{sm} - \varepsilon_{cm}$

$F_s = M_{Ed}/z$ = 125.5 / 0.299 = 420 kN

σ_s = 420 × 10³ / 3275 = 128 MPa

k_t = 0.4 (long-term loading)

$f_{ct,eff} = f_{ctm}$ = 2.8 MPa

$h_{c,eff}$ = (400 – 144) / 3 = 85 mm

$\rho_{p,eff}$ = 3275 / (85 × 1 000) = 0.039

BS EN 1992-1-1 Eqn (7.9)

$\varepsilon_{sm}-\varepsilon_{cm}$ = $[128 - 0.4.(2.8/0.039).(1 + (32).(0.039)] / 200\,000$
= 0.00032

(Minimum value = $0.6 \times 128 / 200\,000$ = 0.00038)

Determine $s_{r,max}$

φ = 25 mm

c = 40 mm

Verify limit of bar spacing

= $5 \times (40 + 25/2) = 263$ mm > actual bar spacing

k_1 = 0.8

k_2 = 0.5

k_3 = 3.4

k_4 = 0.425

BS EN 1992-1-1 Eqn (7.11)

$s_{r,max}$ = $3.4.(40) + 0.8.(0.5).0.425.(25)/0.039 = 245$ mm

BS EN 1992-1-1 Eqn (7.8)

Determine w_k

w_k = $s_{r,max} \cdot (\varepsilon_{sm}-\varepsilon_{cm}) = 245 \times 0.00032 = 0.08$ mm

w_{k1} = 0.18 mm

$\boldsymbol{w_k < w_{k1}}$

Flexural crack width satisfactory for B25 bars @ 150 mm spacing

(Note. Cracking due to thermal and shrinkage effects must also be verified using the methodology in §8. This may affect the reinforcement required.)

Stress limitations

BS EN 1992-1-1 §7.2(3)

Verify concrete and steel stresses for worst-case bending moment in short span direction at mid-span.

Concrete.

Limiting compressive stress

= $k_2 \times f_{ck}$ = 0.45×28 = 12.6 MPa

F_c = F_s = 420 kN

σ_c = $420 \times 10^3 / (0.5 \times 144 \times 1\,000)$ = 5.8 MPa
< limiting stress

Reinforcing steel.

Limiting tensile stress

= $k_3 \times f_{yk}$ = 0.8×500 = 400 MPa

As previously calculated, σ_s = 128 MPa
< limiting stress

Limiting stresses satisfactory

Deflection

BS EN 1992-1-1 §7.4.2 (2)

Verify acceptability of horizontal deflection of slab panel by consideration of the limiting span/depth ratio.

Consider the main flexural reinforcement at mid-span as <u>B25 bars @ 150 mm spacing.</u>

A_s = 3275 mm²/m

b = 1 000 mm

d = 347 mm

ρ_0 = $\sqrt{28} \times 10^{-3}$ = 0.0054

ρ = 3 275/(1 000 × 347) = 0.0094

$\rho > \rho_0$

so use Eq. (7.16.b) to determine limiting l/d ratio.

BS EN 1992-1-1 Eqn (7.16.b)

K = 1.3

ρ' = 0

Limiting l/d ratio

= $1.3 \times [11 + 1.5 \times \sqrt{28} \times (0.0054/(0.0094 - 0)) + 0]$

= 20.2

l/d = 5.300 / 0.347 = 15.2 <3 limiting ratio

<u>Deflection satisfactory</u>

<u>Ultimate Limit State (STR)</u>

Bending

Verify main reinforcement for worst-case bending moment in short span direction at mid-span.

Sheet PH/18

$M_{Ed} = M_{sx}$ = 181.6 kNm/m (hogging)

Verify for <u>B25 bars @ 150 mm spacing</u>

A_s = 3275 mm²/m

b = 1 000 mm

d = 347 mm

f_{ck} = 28 MPa

γ_c = 1.50

f_{yk} = 500 MPa

γ_s = 1.15

Use rectangular stress block

$k = M_{Ed} / (b \times d^2 \times f_{ck})$

$= 181.6 \times 10^6/(1\,000 \times 347^2 \times 28) = 0.054$

Lever arm factor

$= 0.5 + \sqrt{(0.25 - k / 1.14)}$

$= 0.5 + \sqrt{(0.25 - 0.054/1.14)} = 0.950$

(Limiting value $= 0.950$)

$z = 0.95 \times 347 = 330$ mm

$A_{s\ reqd} = M_{Ed} / (f_{yd} \times k)$

$= M_{Ed} / ((1/\gamma_s) \times f_{yk} \times z)$

$= 181.6 \times 10^6 / (0.87 \times 500 \times 330)$

$= 1265$ mm^2/m

$A_{s\ prov} = 3275$ mm^2/m

$\boldsymbol{A_{s\ prov} > A_{s\ reqd}}$

Bending satisfactory for B25 bars @ 150 mm spacing

Shear

Verify shear in slab along junctions with Walls B and C. Consider a 1 m wide strip of slab spanning between Wall B and Wall C with uniformly distributed loading from bearing pressure.

$V_{Ed} = 0.5 \times (w \times L)$

Sheet PH/16 — $w = 72.9$ kN/m^2

$L = 5.0$ m

$V_{Ed} = 0.5 \times 72.9 \times 5.0 = 183$ kN/m

Verify shear capacity for main tension reinforcement at support, i.e. B20 bars @ 150 mm spacing ($A_s = 2095$ mm^2/m).

BS EN 1992-1-1 §6.2.2 — Determine $V_{Rd,c}$

$C_{Rd,c} = 0.18 / \gamma_c$

$= 0.18 / 1.50 = 0.12$

$k = 1 + \sqrt{(200 / 347)}$

$= 1.76$

$\rho_l = 2095 / (1\,000 \times 347)$

$= 0.006 <$ limiting value (0.02)

$V_{Rd,c} = [0.12 \times 1.76 \times (100 \times 0.006 \times 28)^{1/3}]$

$\times 1\,000 \times 350 \times 10^{-3}$

$= 189$ kN/m

$\boldsymbol{V_{Rd,c} > V_{Ed}}$

No design shear reinforcement required

BS EN 1992-1-1 §6.2.3(7)

Check additional tensile force in longitudinal reinforcement due to shear.

$\Delta F_{td} \leq 0.5 \times 183 \times (2.5 - 0) = 229$ kN

$A_{s\ reqd} = 229 \times 10^3 / (0.87 \times 500) = 526$ mm²/m

Area of bending reinforcement required at slab/wall junction (Wall C, as for Wall A)

PH/24

$= 905$ mm²/m

Total area of longitudinal reinforcement required (bending + shear)

$= 905 + 526$

$= 1431$ mm²/m

$A_{s\ prov} = 2095$ mm²/m

$\boldsymbol{A}_{\mathbf{s\ prov}} > \boldsymbol{A}_{\mathbf{s\ reqd}}$

B20 bars @ 150 mm spacing satisfactory

Summary

Main reinforcement in the short-span direction is verified as follows:

- **B25 bars @ 150 mm spacing (Top) at mid-span**
- **B20 bars @ 150 mm spacing (Bottom) at support**

Bending in long span direction at mid-span

Serviceability Limit State (SLS)

Sheet PH/19

$M_{Ed} = M_{sy} = 43.8$ kNm/m (hogging)

Ultimate Limit State (STR)

Sheet PH/18

$M_{Ed} = M_{sy} = 73.3$ kNm/m (hogging)

Following the same procedures as used for the short-span direction, B16 bars @ 150 mm spacing can be verified as being satisfactory for SLS and ULS(STR) limit states

Summary

Main reinforcement in the long-span direction is verified as follows:

- **B16 bars @ 150 mm spacing (Top and Bottom)**

9. EARLY-AGE AND LONG-TERM THERMAL AND SHRINKAGE EFFECTS

Introduction

BS EN 1992-3

Determine the effects of cracking due to the restraint of early-age thermal effects and long-term thermal and shrinkage effects in accordance with BS EN 1992-3.
As an example of the procedure, the junction between Wall A and the base slab to the Wet Well will be considered.

The calculations presented here, follow the pattern of 'Example 1' in §5.1 of CIRIA publication C660, *Early-age Thermal Crack Control in Concrete,* by P.B. Bamforth. (References to this publication are annotated 'CIRIA C660'.)

(Note: Following common practice, cracking due to the restraint of early-age thermal effects and long-term thermal and shrinkage effects is considered separately from that cracking due to structural loading.)

Basic design data

Sheet PH/1

The concrete mix considered in Section 1, above, will be assumed.

Strength class	C28/35	(f_{ck} = 28 MPa)
Cement content	320 kg/m^3	
Cement type	CIIB-V + SR	(PC + 35% PFA)
Min. nominal cover	40 mm	(cast against formwork)
Aggregate type	Not known	

Sheet PH/20

As previously determined, the limiting crack width for Wall A under Tightness Class 1 is

w_{k1} = 0.17 mm

Calculation of thermal & shrinkage effects

BS EN 1992-3
§M.2 (b)

Calculate crack widths using BS EN 1991-1-1, Expression 7.8 taking,

$(\varepsilon_{sm} - \varepsilon_{cm}) = R_{ax} \cdot \varepsilon_{free}$

CIRIA C660
Eqn 3.1
Eqn 3.2

Restrained strain,

$\varepsilon_r = R_{ax} \cdot \varepsilon_{free}$

$= K_1 \{[\alpha_c.T_1 + \varepsilon_{ca}].R_1 + \alpha_c.T_2.R_2 + \varepsilon_{cd}.R_3\}$

Early-age thermal & shrinkage effects

Determine effects at concrete age of 3 days.

CIRIA C660
Fig.4.7

Temperature drop, T_1

For 300 mm thick concrete section (Wall A) with 35% PFA cast against 18 mm thick plywood.

T_1 = 18°C

Reference	Calculation	
CIRIA C660 §4.5	*Coefficient of thermal expansion, α_c* BS EN 1992-1-1, §3.1.3(5) states that unless more accurate information is available α_c should be taken as 10 microstrain/°C. For materials used in the UK a more realistic value will be used of α_c = 12 microstrain/°C	
CIRIA C660 §4.6.1	*Autogenous shrinkage, ε_{ca}* Obtain value for C28/35 concrete at 3 days. $\varepsilon_{ca(\infty)} = 2.5 \times (28 - 10) = 45$ microstrain $\beta_{as}(3) = 1 - \exp\{-0.2.(3)^{0.5}\} = 0.293$ $\varepsilon_{ca(3)} = \beta_{as}(3).\varepsilon_{ca(\infty)}$ $= 0.293 \times 45 = 13$ microstrain	
CIRIA C660 §4.7.2	*Restraint to movement at joint, R_j* Consider restraint to movement at horizontal junction of 300 mm thick Wall A cast on top of 400 mm thick base slab. h_n = 300 mm, h_o = 400 mm Assume, $A_n/A_o = h_n/h_o = 300/400 = 0.75$ $E_n/E_o = 0.7$ (conservative/rapid cooling) $R_1 = 1 / (1 + (A_n/A_o) \times (E_n/E_o))$ $= 1 / (1 + 0.75 \times 0.70) = 0.66$	
CIRIA C660 §4.9.1	*Creep coefficient, K_1* $K_1 = 0.65$	
CIRIA C660 Eqn 3.2	*Early-age restrained strain, ε_r* For early-age effects consider first term of Eq. (3.2). $\varepsilon_r = K_1.[\alpha_c.T_1 + \varepsilon_{ca}].R_1$ $= 0.65 \times [12 \times 18 + 13] \times 0.66 = 98$ microstrain	
CIRIA C660 §4.8 Table 4.11	*Tensile strain capacity of concrete, ε_{ctu}* As type of aggregate is not known, take value for quartzite from Table 4.11. Modify for concrete of strength class 28/35. $\varepsilon_{ctu} = 76 \times [0.63 + (35/100)] = 74$ microstrain	
CIRIA C660 Eqn 3.3	*Test for early-age cracking* $\varepsilon_r > \varepsilon_{ctu}$ **Cracking predicted**	
CIRIA C660 Eqn 3.5	*Early-age crack-inducing strian, ε_{cr}* $\varepsilon_{cr} = \varepsilon_r - 0.5.\varepsilon_{ctu}$ $= 98 - 0.5 \times 74 = 61$ microstrain	

Reference	Calculation
CIRIA C660 Eqn 3.12	*Minimum area of reinforcement per face of wall,* $A_{s,min}$ Use Eqn.3.12, which is based on BS EN 1991-1-1, Eqn.7.1.
CIRIA C660 Table 3.1	k_c = 1.0 k = 1.0 h = 300/2 = 150 mm (as $A_{s,min}$ is per face of wall)
BS EN 1991-1-1 Eqn 3.4	Calculate $f_{ctm}(3)$ from BS EN 1991-1-1, Eqn.3.4. s = 0.25 for cement class N (pfa > 20%) $\beta_{cc}(3)$ = $\exp\{0.25[1 - (28/3)^{0.5}]\}$ =0.60 f_{ctm} = 2.8 MPa $f_{ctm}(3)$= 0.60 × 2.8 = 1.68 MPa f_{yk} = 500 MPa A_{ct} = 150 × 1 000 = 150 000 mm²/m/face
CIRIA C660 Eqn 3.12	$A_{s,min}$ = 1^2.150 000.(1.68/500) = <u>504 mm²/m/face</u> As previously calculated, for flexural considerations $A_{s,min}$ = 0.5 × 600 = 300 mm²/m/face From practical considerations, provide minimum horizontal reinforcement to walls of <u>B16 bars @ 150 mm spacing</u> in each face. A_s = 1341 mm²/m/face
CIRIA C660 Eqn 3.13	*Crack spacing,* $s_{r,max}$ c = 40 mm φ = 16 mm k_1 = 1.14 (rec. for evaluation of early-age effects) $h_{c,ef}$ = 2.5.(40 + 16/2) = 120 mm $A_{c,eff}$ = 120 × 1 000 = 120 000 mm²/m $\rho_{p,eff}$ = 1 341 / 120 000 = 0.011 $s_{r,max}$ = 3.4.(40) + 0.425.(1.14 × 16)/0.011 = 841 mm
BS EN 1991-1-1 Eqn 7.8	*Early-age crack width,* w_k w_k = $s_{r,max} \cdot \varepsilon_{cr}$ = $841 \times 61 \times 10^{-6}$ = 0.05 mm w_{k1} = 0.17 mm $\boldsymbol{w_{k1} > w_k}$ **<u>Early-age crack width satisfactory for B16 bars @ 150 mm spacing</u>**

Long-term thermal & shrinkage effects

The restraint conditions of the wall/slab junction will be the same for long-term and short-term effects. Long-term thermal and shrinkage effects will act on the pattern of cracking caused by short-term effects.

CIRIA C660 §4.3

Long-term temperature change, T_2

Assume concrete cast in summer (conservative).

$T_2 = 20°C$

CIRIA C660 §4.6.1

Autogenous shrinkage, ε_{ca}

Obtain value for C28/35 concrete at 28 days.

$\varepsilon_{ca(\infty)} = 2.5.(28 - 10) = 45$ microstrain

$\beta_{as}(28) = 1 - \exp\{-0.2.(28)^{0.5}\} = 0.653$

$\varepsilon_{ca}(3) = \beta_{as}(3).\varepsilon_{ca(\infty)} = 0.653 \times 45 = 29$ microstrain

Drying shrinkage, ε_{cd}

Calculate $\varepsilon_{cd,0}$ using BS EN 1991-1-1, Appendix B.2.

CIRIA C660 §5.1

Assume mean relative humidity of 90%.

BS EN 1991-1-1 Appendix B.2

$\beta_{RH} = 1.55.[1 - (90/100)^3] = 0.42$

$f_{cm} = 36$ MPa

For cement class N (pfa > 20%),

$\alpha_{ds1} = 4, \quad \alpha_{ds2} = 0.12$

BS EN 1991-1-1 Eqn B.12

$\varepsilon_{cd,0} = 0.85.[(220 + 110 \times 4).\exp(-0.12 \times 36/10)].0.42$

$= 153$ microstrain

Calculate ε_{cd} using BS EN 1991-1-1, Eqn.3.9.

$k_h = 0.75$ (300 mm thick section drying from one side)

For large values of 't' (i.e. 30 years/10 950 days), $\beta_{ds}(t,t_s) \rightarrow 1$

BS EN 1991-1-1 Eqn 3.9

$\varepsilon_{cd} = 1 \times 0.75 \times 153 = 115$ microstrain

Restraint to movement at joint, R_j

$R_1 = R_2 = R_3 = 0.66$

CIRIA C660 §4.8 Table 4.11

Tensile strain capacity of concrete, ε_{ctu}

As type of aggregate is not known take value for quartzite from Table 4.11.

Modify for concrete of strength class 28/35.

$\varepsilon_{ctu} = 108.[0.63 + (35/100)] = 106$ microstrain

CIRIA C660
Eqn 3.6

Total crack-inducing strian (early-age & long-term), ε_{cr}

$$\varepsilon_{cr} = K_{1.}\{[\alpha_c.T_1 + \varepsilon_{ca}].R_1 + \alpha_c.T_2.R_2 + \varepsilon_{cd}.R_3\} - 0.5.\varepsilon_{ctu}$$

$$= 0.65.\{[12 \times 18 + 29].0.66 + 12 \times 20 \times 0.66 + 115 \times 0.66\} - 0.5 \times 106$$

$$= 0.65.\{162 + 158 + 76\} - 53 = 205 \text{ microstrain}$$

BS EN 1991-1-1
Eqn 7.8

Long-term crack width, w_k

$$w_k = s_{r,max} \,.\, \varepsilon_{cr}$$

$$= 841 \times 205 \times 10^{-6}$$

$$= 0.17 \text{ mm}$$

$$w_{k1} = 0.17 \text{ mm}$$

$$\boldsymbol{w_{k1} = w_k}$$

Long-term crack width satisfactory for B16 bars @ 150 mm spacing

6.1.5 Indicative Reinforcement Details

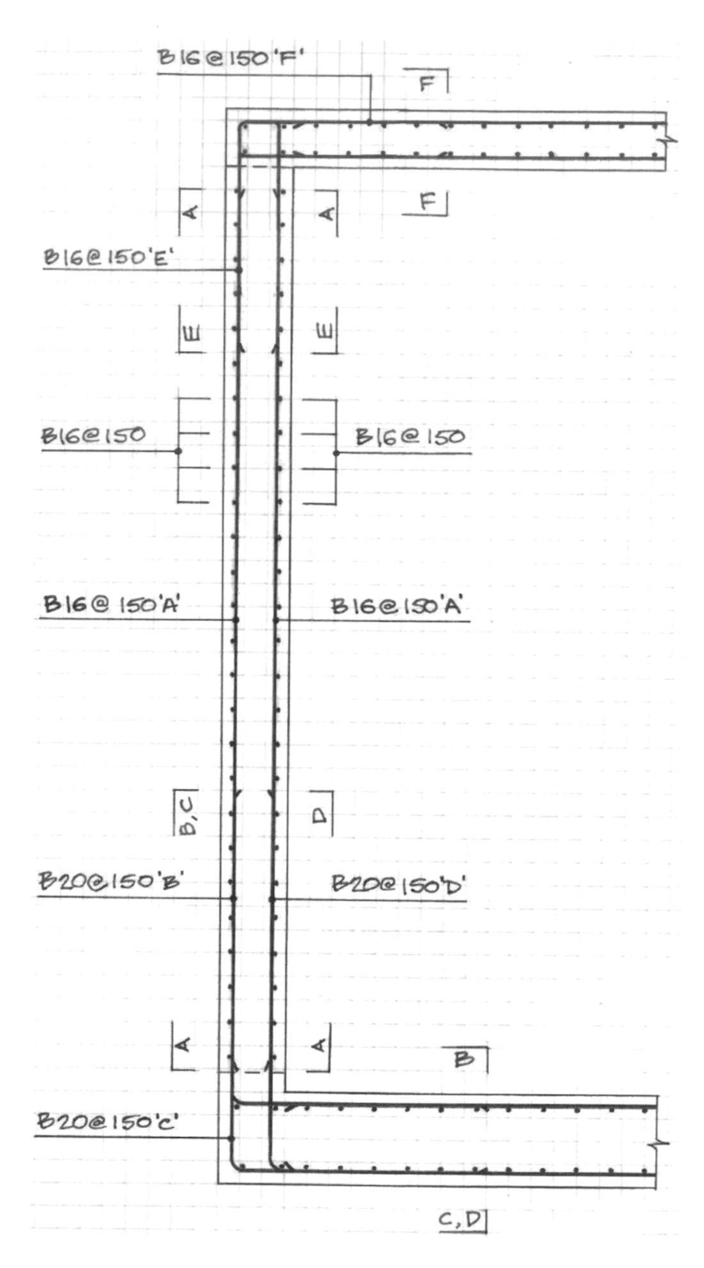

Figure 6.3 *Indicative reinforcement details - Wall A.*

Chapter 7
Testing and rectification

7.1 Testing for watertightness

The design and construction of liquid-retaining structures require close attention to detail by both the designer and contractor, and in spite of the best intentions of both parties, errors and omissions can occur. Equally, random occurrences, unfavourable statistical conjunctions and insufficient design guidance can result in a structure that is less than completely liquid-tight. It is therefore necessary to test the structure after completion to ensure that it is satisfactory and that it complies with the specification.

The method of test depends on the visibility and position of the elements of the structure. The walls of overground structures can be inspected for leaks on the outer face and, if the walls are finally to be backfilled with soil, the inspection can be made before the fill is placed. The walls of underground structures can be inspected if there is sufficient working space available (compliance with the HSE Confined Spaces Regulations 1997 (2009) must be achieved). The floor slabs of all structures built on soil cannot be inspected for leaks, and other methods of test have to be used (CESWI 7th Ed., 2011). The floor of an elevated reservoir (or water tower) can be inspected in the same way as walls, as can the underside of a flat reservoir roof. If the structure is designed to exclude rather than retain water, it is possible to inspect the inside faces of the walls and floors but rarely possible to ensure that liquid is available to make a test, and hardly ever possible to take remedial measures from the outside. An example of this situation is that of a basement of a building that is designed to exclude groundwater. Detailed methods of testing are described in the following sections.

7.2 Definition of watertightness

The term 'watertightness' although descriptive is not sufficiently precise for the purposes of a contract specification. Essentially, a watertight structure is built to contain (or exclude) a liquid, but some loss of liquid is inevitable due to evaporation or slow diffusion through the concrete. Also, actual leaks may occur through fine cracks in the concrete. As mentioned previously in Chapter 3 these may heal autogenously (i.e. without any treatment); part of this mechanism involves water percolating through the crack and dissolving calcium salts from the cement. As the process continues, the crack is slowly filled and eventually the water penetration ceases. The process may take up to about one week with cracks of 0.1 mm width, but up to 3 weeks for cracks of up to 0.2 mm width. Cracks over 0.2 mm thick may not self-seal at all. (BS EN 1992-3 Clause (113) states that 'in the absence of more reliable information, healing may be assumed where the expected range of strain at a section under service conditions is less than 150 microstrain.') The result of autogenous healing is a white excrescence along the line of the crack but no further loss of liquid. This may be acceptable as a permanent feature in some types of structure such as underground tanks but could not be allowed in the walls of an elevated water tower. A further form of leakage consists of damp patches on the surface of a wall. The liquid flow is very small, but the appearance may not be acceptable.

It is essential that the required standard of watertightness is clearly described in the contract specification so that there is no misunderstanding about the quality of result required from the contractor. Watertightness is mentioned in Section 7.3.1 and Table 7.105 of BS EN 1992-3 and related to crack width. It is also discussed in Chapter 4.

7.3 Water tests

A completed structure may be tested by filling with water and measuring the level over a period of time. The concrete in the structure must be allowed to attain its design strength before testing commences, and all outlets must be sealed to prevent loss of water through pipes, overflows and other connections. The structure should also be cleaned.

The structure is slowly filled to its normal maximum operating level. If the structure is filled too quickly, the sudden increase in pressure is likely to cause cracking. As a guide, a swimming pool or relatively small tank could be filled over a period of 3 days, but a large reservoir will take much longer to fill because of the volume of water required. No guidance is available in BS EN 1992-3 with respect to filling rates; the authors refer the reader to the superseded BS 8007, which limits the rate of filling to a uniform rate of not greater than 2 m in 24 hours.

To allow the concrete to become completely saturated with water, a stabilising period is allowed after filling has been completed. The length of the stabilising period depends on the design surface crack width and hence the time required to complete any autogenous healing that may be necessary. For a design crack width of 0.1 mm, a period of one week may be required, but for a design crack width of 0.2 mm, a period of up to 3 weeks is necessary. These times may be adjusted as appropriate. If it is obvious that there is no leakage through cracks after some days, it may be possible to commence the record test somewhat earlier. At the commencement of the test, the level of the water is recorded, and subsequently each day for a further period of 7 days. The difference in level over the period of 7 days is then used to assess the result of the test. The levels may be measured by fixing scales to the walls, or by making marks on the walls above the water line and measuring down to water level with a moveable scale or other device. The level should be recorded at four positions but with a large reservoir, at eight to twelve positions to guard against errors in reading and local settlements.

An open structure (or a closed structure where the air above the water is affected by wind movements) may lose moisture by evaporation, or may gain water due to rainfall. In assessing the results of the water-level readings during the test, allowance must be made for these variations. A simple method of achieving this is to moor watertight containers 80% filled with water at points on the water surface. The water surface inside the container is subject to the same gains and losses as the water in the main reservoir. By taking measurements (x) of the water level in the container from the top edge of the container, the gains or losses due to rainfall and evaporation in the main reservoir may be assessed (Figure 7.1).

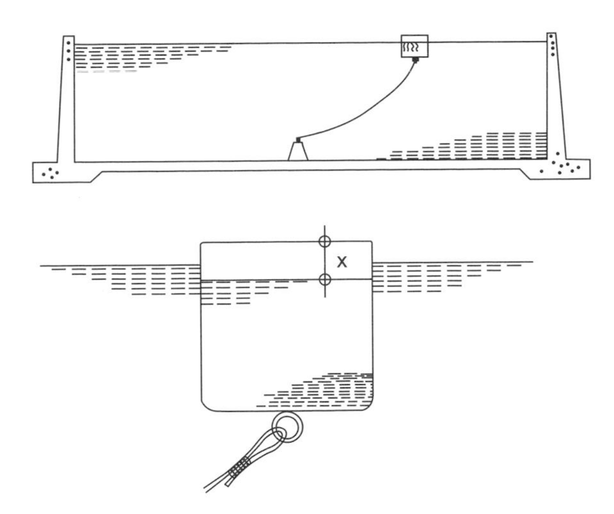

Figure 7.1 *Arrangements for water test.*

It will be apparent that a degree of honesty and care is necessary when carrying out tests of this nature, and the daily measurement of water levels during the test will assist in detecting any unusual occurrences. The authors have had experience of a test where the water level was appreciably higher at the end of the test than at the beginning.

7.4 Acceptance

A water test will enable a net loss of water to be measured due to leakage and further absorption into the concrete structure. The acceptable fall in water level should be stipulated by the designer before the test is commenced. For many structures, the maximum acceptable limit may be taken as 1/1 000 of the average depth of the water. BS 8007 recommends a value of 1/500 of the average water depth or 10 mm or other specified amount; this is still considered acceptable by the authors. It is not possible to set a limit less than about 3 mm due to the difficulty of making a sufficiently accurate measurement.

If the test is judged to be unsatisfactory after 7 days, and if the daily readings indicate that the rate of loss of water is reducing, the designer may decide to extend the test period by a further 7 days. If the net loss of water is then no greater than the specified value during the second period of 7 days, the test may be considered satisfactory. If the test is judged to be unsatisfactory, then it should be repeated after measures have been taken to locate and deal with the leakage.

A successful water-level test is a necessary but not sufficient criterion for accepting the structure as satisfactory. If seepage can be observed from the 'dry' side of the walls or if damp patches are present, then remedial work will be required. The condition of the surface of elements of the structure should be assessed by reference to the contract specification. The structure should not be accepted as satisfactory until the specification has been satisfied in each particular. BS EN 1992-3 provides no guidance on acceptance criteria except to say that acceptance of structures may include the monitoring of the maximum level of leakage (Section 7.3.1 Cl (117)).

7.5 Remedial treatment

It is particularly difficult to isolate defects in floor slabs, but joints and areas where the concrete surface is irregular or honeycombed should be inspected very thoroughly. When water has been drained out of a structure, and the surface is drying, areas containing defects may be the last parts to remain wet or damp, due to water being trapped in the defective area.

Small leaks and damp patches are usually self-healing after 2 to 3 weeks. After the healing is complete, accretions on the outside of the leak may be scraped off the surface. More persistent leaks require treatment with proprietary products, preferably from the water face. Chemicals are available that are applied to cracks as a slurry and are drawn into the crack by the water flow. Fine crystals are formed that close the crack. A similar effect occurs when the slurry is applied to a porous area. Areas of severe honeycombing or wide cracks may be repaired with pressure grouting techniques or, if there is severe leakage, a whole section may need to be cut out and replaced.

Chapter 8
Vapour exclusion

8.1 The problem

There is a trend towards ever larger buildings in city centres, and because of a shortage of land, there are frequently one or more basements below ground level. The basements may be used for car parking, storage, or as office or shop accommodation. A recent phenomenon has been the almost ubiquitous slow rise in groundwater levels most likely as a result of climate change. In many cities in the UK, this rise has been compounded by a reduction in pumping for industrial uses.

The previous chapters in this book have addressed the design of structures to retain aqueous liquids, and by using the same principles, it is possible to design structures to exclude groundwater–as in a basement situation. When a structure is designed to BS EN 1992-3 to exclude liquid, it is accepted that there may be a damp patch or two on the walls, but there should be no leakage of water. A basement in this condition will be acceptable for use as a car park, but for use as a retail sales floor, passage of water vapour (or damp) must be prevented. Although the scope of BS EN 1992-3 covers 'liquid tightness' it does not provide any design guidance with respect to the migration of vapour through the elements of the structure. Properly compacted concrete will prevent the passage of water but will still allow water vapour to migrate through the structural elements, particularly if the basement is heated and/or ventilated. For reasons of health, the UK Building Regulations require all habitable rooms to be designed so that vapour may not pass through the external enclosure below ground level. If water vapour is to be excluded, then additional measures are necessary. BS 8102:2009 provides guidance on the protection of structures against water from the ground and, in particular, prevention from the migration of water vapour. (Note: BS 8102:2009 is a revised version of BS8102:1990. The latest version of the code addresses recent developments such as deeper construction in urban areas, the increase in residential basements, the improvement in waterproofing materials and a more detailed assessment of the potential risks inherent in this type of construction and their mitigation. The designer should be aware that previous guidance available in the 1990 version of the code has been excluded from the 2009 version.) Guidance on water vapour exclusion may also be sought from manufacturers' information sheets.

The following sections deal with the ways in which a vapour-excluding concrete structure may be achieved. It is not possible in this book to deal with all the details applicable to particular materials, and to stay within the limits of the subject of this book, it is assumed that the underground structure is of new in-situ reinforced concrete construction.

8.2 Design requirements

Although it may seem that a vapour-excluding basement is sufficiently well-defined by its description, this is not so. Before design commences, discussion is necessary between the designer and the building owner to decide on the level of protection required, i.e. the grade of basement. (BS 8102:2009 provides a Design Flowchart–Figure 1.) There are three grades of basement: Grade 1 (Basic Utility), Grade 2 (Better Utility) and Grade 3 (Habitable). Grade 4 (Special) is no longer referenced in BS 8102:2009 (archive storage is referenced in BS 5454, 2000). A balance must be struck between satisfactory performance of the structure in use and the cost of providing the protection. The consequences of failure and the anticipated life of the building (or contents) are also part of the considerations. It is extremely important to establish the position of the water table and its potential variability to mitigate potential risks to the final construction. Once the basement grade has been established, the partners need to decide on the type of basement construction. This second stage can be achieved in accordance with BS 8102:2009.

BS 8102:2009 defines the levels of protection required for various specified uses in terms of three types of construction, i.e. A, B and C. Table 8.1, which is based on Table 1 and 2 of BS 8102 and Table 2.2 of CIRIA Report R140,1995 summarises the grade of basement and the types of construction (together with the usages and corresponding performance levels).

To satisfy each of these criteria, a structure may be designed in several ways, each being designated by the type letter, as follows:

Table 8.1 *Guide to level of protection and basement use.*

Grade	*Basement usage*	*Performance level*	*Form of construction*
1	Car parking; plant rooms (excluding electrical equipment); workshops	Some seepage and damp patches tolerable	Type B. Reinforced concrete design in accordance with BS EN 1992-1-1
2	Workshops and plant rooms requiring drier environment; retail storage tolerable areas	No water penetration but moisture vapour	Type A. Type B. Reinforced concrete design in accordance with BS EN 1992-3.
3	Ventilated residential and working areas including offices, restaurants etc., leisure centres	Dry environment	Type A. Type B. With reinforced concrete design to BS EN 1992-3. Type C. With wall and floor cavity and DPM

Type A. With reinforced concrete design to BS EN 1992-1-1 plus a vapour-proof membrane.
Type B. With reinforced concrete design to BS EN 1992-3 plus a vapour-proof membrane.
Type C. With ventilated wall cavity with vapour barrier to inner skin and floor cavity with DPM.

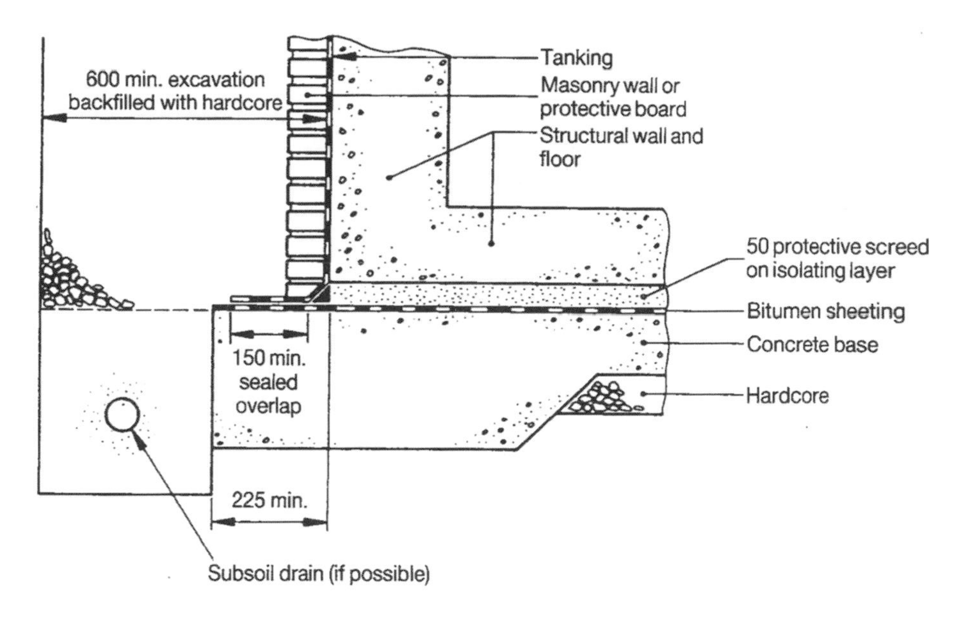

Figure 8.1 *External membrane protection.*

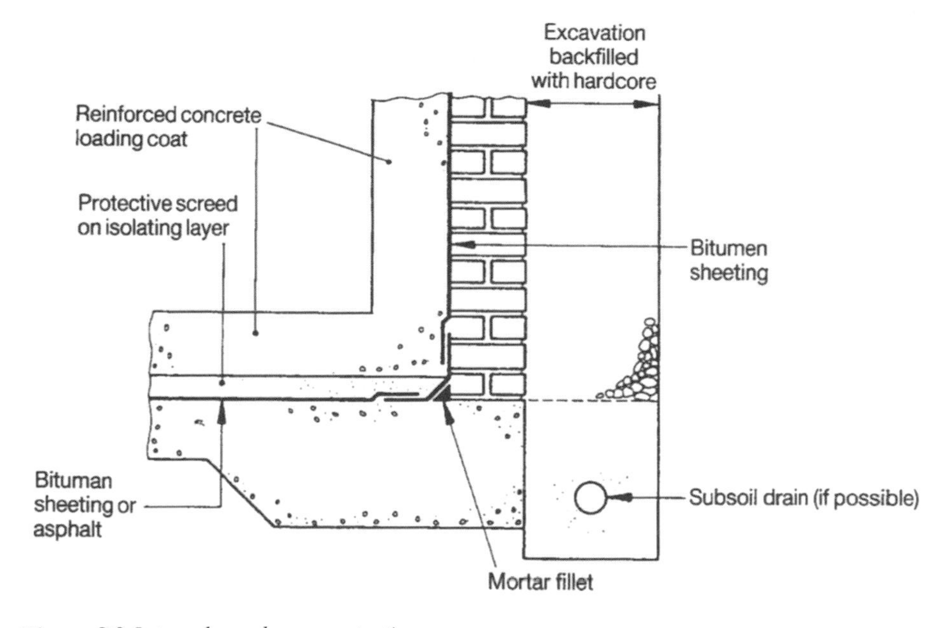

Figure 8.2 *Internal membrane protection.*

Type A–Barrier or tanked protection

The structure itself is not water excluding (it will be designed to the requirements of BS EN 1992-1-1 not BS EN 1992-3) and protection is provided by a membrane system applied either externally or internally. The tanking may either exclude both water and vapour or be only water excluding.

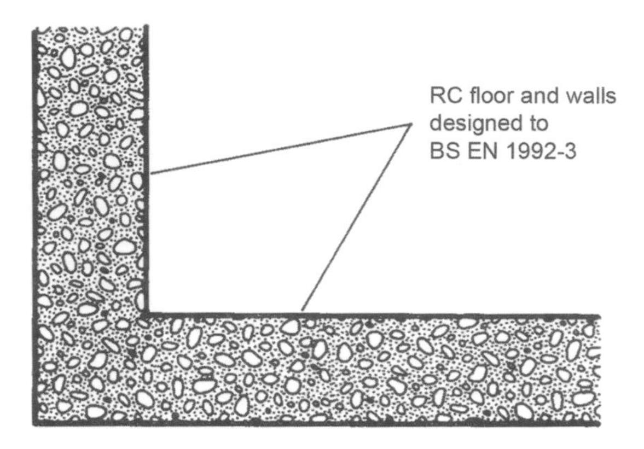

Figure 8.3 *Structurally integral protection.*

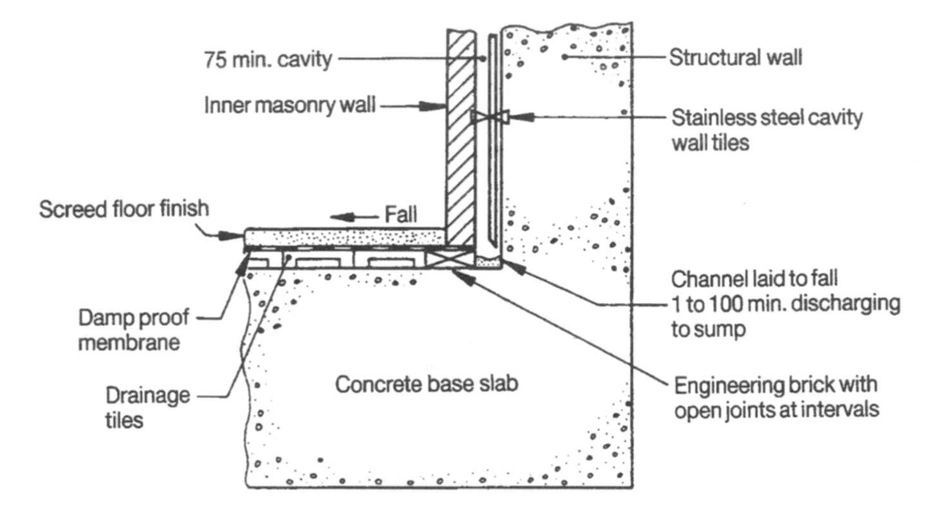

Figure 8.4 *Drained cavity construction.*

Type B–Structurally integral protection

The structure is designed to BS EN 1992-3 to be water excluding but will not be vapour excluding unless an external or internal membrane system is applied (i.e. may require Type A or C protection). The structure can be designed to BS EN 1992-1-1 if the designer only needs to minimise water penetration.

Type C–Drained protection

A drained and possibly ventilated cavity wall construction is provided together with drained cavity floor construction. The floor construction also includes a damp-excluding membrane. The external wall should still be designed to BS EN 1992-1-1 to prevent ingress of significant amounts of water.

Figures 8.1–8.4 show examples of these types of construction, and Table 8.1 includes recommendations of their application for the various grades of protection required.

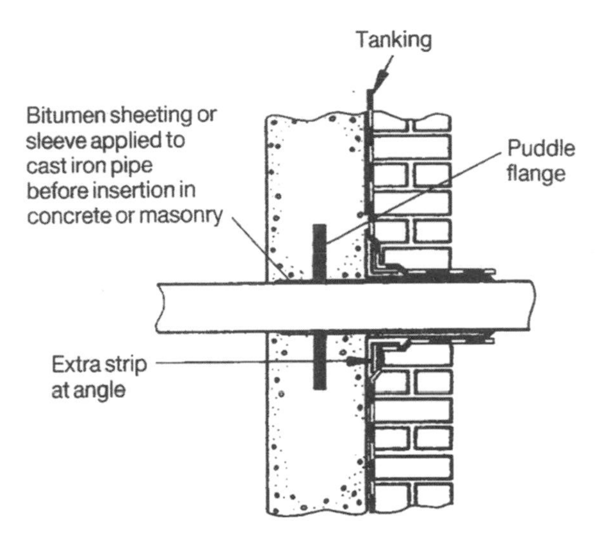

Figure 8.5 *Typical detail of service entry.*

As indicated in the definitions above, more than one type of construction is available for each level of protection. To reiterate, it is not possible to 'design' reinforced concrete to prevent the passage of vapour, and hence an additional barrier of an appropriate material is necessary. The essential feature of the barrier is that it should be continuous, with particular attention given to the junction between floor and walls and to the effective sealing of any pipes or services that pass through the walls or floor (see Figure 8.5).

8.3 Assessment of site conditions

The water and water vapour that are to be excluded from a basement come from groundwater, local surface water, or fractured water supply or drainage pipes. It is important to provide protection from rain falling on the surfaces adjacent to the building, and paved areas should be provided around the structure that will allow surface water to be drained away.

In the 1990 version of BS 8102, it was recommended that in the design of basements not exceeding 4.0 m in depth, particularly when considering stability (uplift/flotation), the design head of groundwater should be assumed to be three-quarters of the full depth of the basement below ground level (but not less than 1 m). And for deeper basements, the water table should be taken as being 1 m below ground level. This may sometimes seem to be a very conservative approach, but it is important to remember that if a basement is excavated in clay soil and backfill is placed around the completed structure, then a sump has been created that will tend to attract any surface water in the vicinity. Although this guidance has not been included in the latest, 2009, version of the code, it is still considered good practice.

A comprehensive soils investigation is necessary for all but very small jobs and, in the case of basement construction, it is important to obtain detailed information concerning any groundwater table together with an indication of the likely variation of that table both seasonally and over the anticipated life of the building. Guidance on what constitutes a detailed Site Evaluation is provided in BS 8102:2009. This includes the determination of ACEC (aggressive chemical environment class) class and DC (design chemical) class according to BRE Special Digest 1 (2005). Information about the quality of the soils and groundwater in terms of pH value and any dissolved chemicals (i.e. sulphates and any other chemicals present from previous uses of the ground) may well influence the design decisions concerning the use of an external or internal membrane (see Figures 8.1 and 8.2).

8.4 Barrier materials

The essential properties for a barrier material are that it should be inherently vapour excluding and that it should be of a form that can be conveniently applied to the main structure. This includes the ability to negotiate corners and changes of level and to remain stable in a vertical application to a wall. The structure onto which the barrier material is placed should not contain uncontrolled cracks that might rupture the material. Hence, design of concrete to BS EN 1992-3 is to be preferred (although in certain cases design to BS EN 1992-1-1 is still acceptable–see Section 8.2). Details for any movement joints should be prepared to preserve the exclusion of vapour, and also at any change in backing material (e.g. brick to concrete). It should be noted that a vapour-excluding barrier or membrane will also prevent water penetration, assuming that the barrier material is not forced away from the structure by water pressure. The main materials in use are described below. To specify each material in detail it is necessary to consult BS 8102:2009 and other appropriate British Standards and manufacturers' literature.

Protection of the material is generally required after it has been placed. This is applied both on the outside of a structure, before backfilling and on the inside of the structure by providing a loading material to prevent vapour pressure blowing the material away from the structure.

8.4.1 Mastic asphalt membranes

Mastic asphalt is a material that has been used widely for many years. It is applied hot (so that it can flow) and worked into position by hand or by mechanical means. It is applied in three coats of 10 mm per coat. In vertical work, it may require support at intervals due to the weight of the material. The joints in each layer are staggered to avoid possible paths for leakage. Where asphalt is applied to the exterior of the structure, it requires protection before backfill is placed.

8.4.2 Bonded sheet membranes

This material consists of a sheeting material coated with bitumen. It is supplied in rolls of various weights and widths, and applied cold (i.e. self-adhesive) or hot (i.e. using a heating gun or bonded using a hot melt bitumen adhesive). The surfaces onto which the material is applied should be smooth and free from rough edges. At least two layers are required, with the lines of the joints being staggered in position.

8.4.3 Cement-based renders

Cementitious crystallisation slurries are mixed on site and are a blend of sand, Portland cement and a waterproofing admixture or a polymer resin. Water is added and the mixture applied in two coats with staggered joints. These renders are not necessarily entirely vapour excluding. It is important to ensure that the backing materials are in a satisfactory condition to receive the render and that the backing is stable and uncracked. Rendering over a change in materials is not likely to be satisfactory as cracks will form in the render over the lines of change. No protection to the render is normally necessary. Cementitious multi-coat renders, mortars and coatings also exist. Application of this type of material should be left preferably until the structure's permanent load has been applied.

8.4.4 Liquid applied membranes

Various products are available that are supplied as a liquid or semi-liquid and are applied by roller, trowel or other means specified by the manufacturer. The resin cures after a period of one to two days forming a jointless vapour-excluding sheet.

8.4.5 Geosynthetic (bentonite) clay liners

These are comprised of bentonite with a single or dual 'carrier' material (i.e. a geotextile or high-density polyethylene). The liners can be applied dry, where the activation relies on the absorption of groundwater once installed. Alternatively, they can be supplied prehydrated.

8.5 Structural problems

As mentioned previously, BS 8102:2009 provides a detailed assessment of the risks inherent in constructing these types of structure and how to mitigate these risks. This section and Section 8.6 highlight a few of the more serious problems that can be encountered by the designer and certain critical aspects that need to be considered. Guidance should also be sought from the Concrete Centre Best Practice Guide (BRE, 2005).

8.5.1 Construction methods

During construction, it is almost always necessary to support the ground outside basement walls, and this has an effect on the construction sequence and the positioning of joints. If groundwater is present at a relatively high level, then sheet piling, diaphragm walling, or a system of well-points may be required. The design must take account of any restrictions that the construction method imposes.

8.5.2 Layout

The layout of the basement structure will be influenced by the method of construction and, in particular, by the means used to support the ground at the sides of the excavation. If temporary sheet piling is used, it is more economic if the junction of the floor and the wall has no heel projecting beyond the outside face of the wall. However, this may conflict with the need for an overlap of the barrier material at the wall / floor junction.

8.5.3 Piled construction

For vapour-excluding structures, construction on piles requires a complete separation between the pile caps with their stabilising beams and the wall and floor structure (Figure 8.6). It occasionally happens that tension piles are required to hold down the

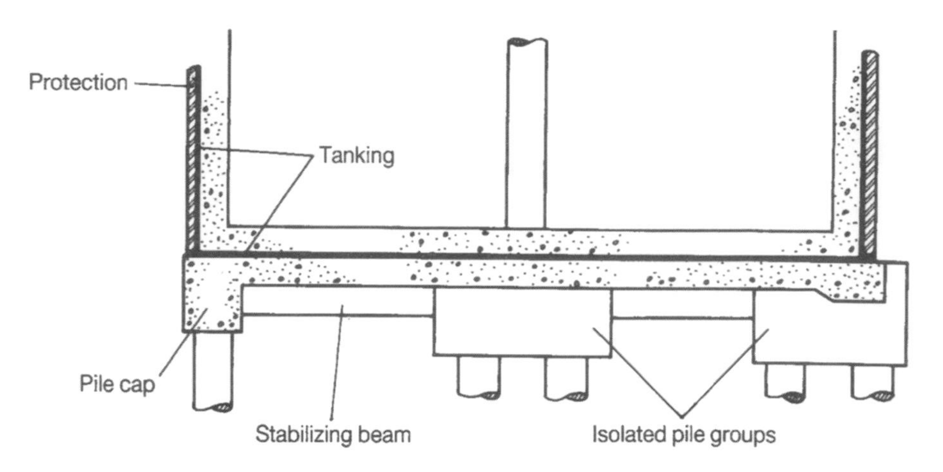

Figure 8.6 *Piled construction.*

basement structure against uplift forces due to groundwater. This creates a particular problem as the tension reinforcement in the piles must be properly anchored in the main basement structure, and yet any membrane must be continuous. The possible solutions are either to devise a special local joint around the tension bars, or to use cavity construction.

8.5.4 Diaphragm and piled walls

The use of diaphragm walls or contiguous piled walls is extremely convenient when an excavation has to be carried out alongside an existing building, but the nature of these systems is such that they cannot be relied upon to be water excluding. The simplest solution to this problem is to use cavity construction (Figure 8.4). This is a system where it is accepted that there will be some penetration of the main structure by vapour and possibly water. A system of lining walls is provided and positioned to form a cavity that separates the main structure from the inner lining. Similarly, a secondary floor is provided, which allows for an air space between the main structural floor and the secondary floor. The floor is provided with a vapour-excluding layer. Arrangements are made so that any water that collects in the cavities can be drained away to a sump and pumped out. Vapour may be removed by ventilating the cavity. The degree of protection required will be determined by the particular use of the building (Table 8.1).

8.6 Site considerations

8.6.1 Workmanship

Although the quality of workmanship is important in all building operations, the construction of vapour-excluding structures demands workmanship of the highest quality. The reasons for this are as follows.

(a) Moisture can easily migrate from a defect behind a membrane to emerge on the opposite face in an entirely different position. The source of any leakage of water or transmission of vapour is difficult to locate.

(b) When an external membrane is used, it is virtually impossible to gain access to the underside of the floor slab or the outer faces of the external walls without enormous cost and disruption.
(c) In general, it is not possible to check that a structure is vapour excluding during the construction phase when there is a great deal of moisture present. Some defects may not be revealed before the heating is activated.

The work involved in the application of membranes to a concrete wall is straightforward, but it requires dedication and detailed care. In adverse weather conditions, work may have to be halted. If the construction sequence requires a section of work to remain in a part-finished state for some time, then the exposed temporary edge may need protection, and the joint between the old and new will require careful treatment by cleaning the previous work before bonding on the new.

It cannot be stressed enough that the failure of the vapour barrier effectively constitutes the failure of the structure. It is imperative that the application of the barrier is performed by skilled personnel and is not seen or treated as a minor task. The consequences of a failed barrier can be significant not only with respect to the performance of the structure but also financially!

8.6.2 Failure

The author has inspected a basement that was to be used as a retail trading floor, and the structure was subjected to groundwater pressure from a level of 800 mm below the surface. The structure was quite correctly designed at the time to BS 8007, with the addition of an externally-applied membrane. In spite of these features, the structure leaked profusely. The workmanship on the application of the membrane was very poor, and the waterstops that had been inserted in the construction joints were ineffective–again due to faulty workmanship.

It is not sufficient for a contractor to hire the next man on the list from the labour exchange and put a brush in his hand. The operatives must be properly trained and preferably have relevant experience. Supervision is also important and needs to be nearly continuous. To execute a design correctly costs money, but the cost of satisfactory repairs will be many times greater.

The importance of using suitably experienced personnel for this task is now stressed in BS8102:2009. An incorrectly installed barrier can lead to serious litigation problems. The designer should be aware that both the contractor and the designer have a duty of care. Owing to the critical nature of this task, the designer has a responsibility to ensure that the barrier is installed correctly. It is not acceptable to leave this to the contractor. On large contracts, it is likely that the designer will be represented on site by a Resident Engineer. However, for small contracts, mitigation of this problem may be that the designer has to attend site when the barrier is installed.

8.6.3 Services

It is frequently required to pass pipes or services through a water and/or vapour-excluding wall. It is preferable to cast service pipe ducts etc. into the wall rather than leave a hole to be made good later. A puddle flange should be provided around pipes etc. at the centre of thickness of the wall. Puddle flanges can be provided on both cast iron and plastic pipes, but with plastic pipes a further problem occurs due to the flexibility of the material. There is a possible lack of adhesion between the surface of the pipe and concrete (leading to leakage). A convenient method of improving the adhesion between plastic pipes and concrete is to paint the outside surface of the pipe with epoxy adhesive and scatter dry sand onto the surface. This technique produces a surface similar to glass paper, and reduces the possibility of any leakage (Figure 8.5).

8.6.4 Fixings

When a basement is used for storage, retail activity or other similar purposes, there will always be a requirement to fix signs, shelves, services and other items to the walls. If the vapour-excluding barrier is placed on the inside of the structural walls, the fixings will penetrate the barrier and destroy its effectiveness. It may be possible to design local details to overcome this problem, but, in general, the original designers or developers of a building will not have control over the activities of the occupants, and eventually the vapour barrier will be compromised. This problem arises irrespective of the material used for the barrier. There is less of a problem when services are required in a floor, as they can be embedded in a screed above the vapour barrier. If any drainage goods are specified in the floor, they should be made of cast iron rather than ceramic or plastic as it is otherwise not possible to make a satisfactory vapour seal around pipes and gullies.

References

ACI Committee 207, Effect of restraint, volume change and reinforcement on cracking of mass concrete, *ACI Materials Journal,* May-June 1990, 87(3) pp. 271–295.

Al Rawi, R. S. and Kheder, G. F., Control of cracking due to volume change in base-restrained concrete members, *ACI Structural Journal,* July–August 1990, 397–405.

Alexander, S., Understanding shrinkage and its effects: part 2. *Concrete,* 36(10), November/December 2002.

Anchor, R.D., *Design of Liquid Retaining Concrete Structures,* Second edition, Edward Arnold, London, 1992.

Bamforth, P. B., CIRIA C660, *Early-age Thermal Crack Control in Concrete,* CIRIA, London, 2007.

Bamforth, P. B., Denton, S. and Shave, J., *The development of a revised unified approach for the design of reinforcement to control cracking in concrete resulting from restrained contraction,* The Institution of Civil Engineers, Research project 0706, Feb. 2010, 67 pages–Final Report.

Barnes, G. E., *Soil Mechanics, Principles and Practice,* Second edition, Macmillan, Basingstoke, 2000.

Base, G. D., Read, J. B., Beeby A. W. and Taylor, H. P. J., *An investigation of the crack control characteristics of various types of bar in reinforced concrete beams,* Research Report 41-018, Cement and Concrete Association, London, 1966.

Beeby, A. W., The prediction of crack widths in hardened concrete, *The Structural Engineer,* January 1979, 57a (1), 9–17.

Beeby, A. W., *Fixings in cracked concrete,* CIRIA Technical Note 136, 1990.

Beeby A. W., EC2 Part 3; Code Committee minutes, 2000-2005–not published.

Beeby, A. W., The influence of the parameter φ/ρeff on crack widths, *Structural Concrete,* 5 (2), June 2004, 71–83.

Beeby, A. W., Discussion of 'The influence of the parameter φ/ρ_{eff} on crack widths', *Structural Concrete,* (4), October 2005, 155–165.

Beeby, A. W., and Forth, J. P., Control of cracking in walls restrained along their base against early thermal movements. In: Dhir, R. K., McCarthy, M. J. and Caliskan, S., *Concrete for Transportation Infrastructure,* Thomas Telford Publishing, 2005.

Beeby, A. W. and Scott, R, H., *Tension Stiffening of Concrete–Behaviour of Tension Zones in Reinforced Concrete Including Time Dependent Effects,* The Concrete Society, Camberley, UK, 2003.

BS EN10025-1:2004 *Hot Rolled Products of Structural Steels.* General Technical Delivery Conditions, British Standards Institution, London, 2004.

BS EN 10080:2007 *Steel for the Reinforcement of Concrete-weldable Reinforcing Steel–General,* British Standards Institution, London, 2007.

BS EN 13877-3:2004 *Concrete Pavements. Specifications for Dowels to be Used in Concrete Pavements,* British Standards Institution, London, 2004.

BS EN 1990 *Eurocode–Basis of Structural Design,* British Standards Institution, London.
BS EN 1991: 2002 *Eurocode 1: Actions on Structures–Parts 1-1 to 1-6,* British Standards Institution, London.
BS EN 1991-1-1 *Eurocode 1–Actions on Structures–Part 1-1:* General Actions, British Standards Institution, London.
BS EN 1991-4 *Eurocode 1–Actions on Structures–Part 4: Silos and Tanks,* British Standards Institution, London.
BS EN 1992-3: 2006 *Eurocode 2: Design of Concrete Structures–Part 3 Liquid Retaining and Containment Structures,* British Standards Institution, London, 2006.
BS EN 1992-1-1: 2004 *Eurocode 2: Design of Concrete Structures–General Rules and Rules for Buildings,* British Standards Institution, London, 2004.
BS EN 1997-1:2004 Eurocode 7: Geotechnical Design–Part 1: General Rules, British Standards Institution, London, 2004.
BS EN 206-1:2000 *Concrete–Part 1. Specification, Performance, Production and Conformity,* British Standards Institution, London, 2006.
BS 4449:2005 *Carbon Steel Bars for the Reinforcement of Concrete,* British Standards Institution, London, 2005.
BS 4483:2005 *Steel fabric for the reinforcement of concrete.* Specification, British Standards Institution, London, 2005.
BS 5328-1:1997 *Concrete. Guide to Specifying Concrete,* British Standards Institution, London, 1997 (superseded).
BS 5337 *Structural Use of Concrete for Retaining Aqueous Liquids,* British Standards Institution, London.
BS 5454:2000 *Recommendations for the Storage and Exhibition of Archival Documents,* British Standards Institution, London, 2000.
BS 5896 Specification for High Tensile Steel Wire and Strand for the Prestressing of Concrete, British Standards Institution, London.
BS 6399-1:1996 *Loading for Buildings. Code of Practice for Dead and Imposed Loads,* British Standards Institution, London, 1996 (superseded).
BS 648:1964 *Schedule of Weights of Building Materials,* British Standards Institution, London, 1964 (superseded).
BS 6722:1986 *Recommendations and Dimensions of Metallic Materials,* British Standards Institution, London, 1986.
BS 8007 *Design of Concrete Structures for Retaining Aqueous Liquids,* British Standards Institution, London.
BS 8102:2009 *Code of Practice for Protection of Below Ground Structures Against Water from the Ground,* British Standards Institution, London, 2009.
BS 8110 *Structural Use of Concrete (Parts 1 to 3),* British Standards Institution, London.
BS 8500-1:2006 *Concrete–Complementary British Standard to BS EN 206-1 Part 1,* British Standards Institution, London, 2006.
BS 8500-2:2006 *Concrete–Complementary British Standard to BS EN 206-1 Part 2,* British Standards Institution, London, 2006.
BS 8666:2005 *Scheduling, Dimensioning, Bending and Cutting of Steel Reinforcement for Concrete,* British Standards Institution, London, 2005.
Building Research Establishment, *Concrete in aggressive ground,* BRE Special Digest 1, Watford, 2005.
CARES, Introduction of British Standard BS 8666:2005, www.ukcares.com (cited 2012).
CIRIA Report R140, Water-resisting basements, CIRIA, 1995.
Collins, M. P., Mitchell, D. and Bentz, E. C., Shear design of concrete structures, *The Structural Engineer,* 86 (10), May 2008, 32–39.

Davies, J. D., Circular tanks on ground subjected to mining subsidence, *Civil Engineering and Public Works Review,* 55, July 1960, 918–920.

Dhir, R., K. Paine, A., and Zheng, L., *Design data for use where low heat cements are used,* DTI Research Contract No. 39/680, CC2257, University of Dundee, Report No. CTU2704, 2004.

Farra, B. and Jaccoud, J. P., *Influence du beton et de l'armature sur la fissuration des structures en beton–Rapport des essais de tirants sous deformation imposee de court duree,* Departement de Geie Civil, École Polytechnique Fédérale de Lausanne, Publication No. 140, November 1993.

Forth, J. P., Chapter7–Edge Restraint, Concrete Society Tech Report No. 67, *Movement, Restraint and Cracking in Concrete Structures,* 2008.

Forth, J. P., *13 years of monitoring of Cropton buried service reservoir,* Internal Report, School of Civil Engineering, University of Leeds, Leeds, 2012.

Forth, J. P., *Verifying the effective modulus approach for Serviceability calculations,* Internal Report, School of Civil Engineering, University of Leeds, Leeds, 2012.

Forth, J. P. and Beeby, A. W., Study of composite behaviour of reinforcement and concrete in tension, *ACI Structural Journal,* Vol. 111, No. 2, March–April 2014, pp. 397–406.

Forth, J. P., Beeby, A. W. and Scott, R., *Shrinkage curvature of cracked reinforced concrete sections: improving the economy of concrete structures through good science,* Application to EPSRC, funded in 2004.

Forth, J. P., Lowe, A. P., Beeby, A. W. and Goodwill, I. M., Solar effects on a partially buried reinforced concrete service reservoir, *The Structural Engineer,* 83 (34/24), December 2005, 39–45.

Forth, J. P., Mu, R., Scott, R., Jones, T. and Beeby, A. W., Verification of shrinkage curvature models in codes for cracked sections, *Procs of the ICE, Structures and Buildings,* Vol. 167, Issue 5, August 2013, pp. 274–284, from http://www.icevirtuallibrary.com/content/article/10.1680/stbu.12.00046.

Gray, W. S. and Manning, G. P., *Concrete Water Towers, Bunkers, Silos and Other Elevated Structures,* Cement and Concrete Association, London, 1973.

Harrison, T. A., CIRIA 91, *Early Age Thermal Crack Control in Concrete,* Second edition, CIRIA, London, 1991.

Health and Safety Executive, *Safe Work in Confined Spaces, Confined Spaces Regulations, 1997,* HSE Books, L101, 2009, 40 pp.

Higgins, L., Forth, J. P., Neville, A., Jones, R. and Hodgson, T., Long-term behaviour of cracked reinforced concrete beams under repeated and static loading conditions, *Engineering Structures,* 56, 2013, pp. 457–465.

Hobbs, D. W., *Shrinkage induced curvature of reinforced concrete members.* Cement and Concrete Association Development Report No. 4, November 1979.

Hughes, B. P., *Limit State Theory for Reinforced Concrete Design,* Second edition, Pitman, 1976.

Hughes, B. P., Early-age concrete crack control–is EC2 right or wrong? *The Structural Engineer,* 86 (15), August 2008, 32–37.

Kaethner, S., Have EC2 cracking rules advanced the mystical art of crack width prediction', *The Structural Engineer,* 89 (19), October 2011, 14–22.

Kheder G. F., A new look at the control of volume change cracking of base restrained concrete walls, *ACI Structural Journal,* May–June, 1997, 260–271.

Kong, K. L., Beeby, A. W., Forth, J. P. and Scott, R. H., Cracking and tension behaviour in reinforced concrete flexural members, *Proceedings of the Institution of Civil Engineers, Structures and Buildings,* June 2007.

Manning, G. P., *Reinforced Concrete Reservoirs and Tanks,* Cement and Concrete Association, London, 1972.

Martin, L. H. and Purkiss, J. A., *Concrete Design to EN 1992,* Second edition, Butterworth Heinemann, Oxford, 2006.

Melerski, E. S., *Design Analysis of Beams, Circular Plates and Cylindrical Tanks on Elastic Foundations,* A. A. Balkema, Rotterdam, 2000, 284 pp.

Moss, R. and Webster, R., EC2 and BS8110 compared, *The Structural Engineer,* 92 (6), March 2004, 33–38.

MPA The Concrete Centre, *Concrete Basements,* The Concrete Centre, London, April 2012.

Muizzu, M., 'Thermal and time-dependent effects on monolithic reinforced concrete roof slab–wall joints', PhD Thesis (Supervisor Forth, J. P.), School of Civil Engineering, University of Leeds, Leeds, 2009.

Narayanan, R. S. and Beeby, A. W., *Designers' Guide to EN 1992-1-1 and EN 1992-1-2: Designers' Guide to the Eurocodes,* Thomas Telford, London, 2005.

Newman, J. and Choo, B. S., *Advanced Concrete Technology,* Elsevier, Oxford, 2003.

Nilsson, M., *Thermal cracking of young concrete,* Licentiate Thesis, Department of Civil and Mining Engineering, Division of Structural Engineering, Luleå University of Technology, Luleå, Sweden, 2000.

Nilsson, M., *Restraint factors and partial coefficients for crack risk analyses of early age concrete structures,* Doctoral Thesis, Department of Civil and Mining Engineering, Division of Structural Engineering, Luleå University of Technology, Luleå, Sweden, 2003.

Nilsson, M., Jonasson, J-E., Emborg, M., Wallin, K. and Elfgren, L., Determination of restraint in early age concrete walls and slabs by a semi-analytical method–Papers 1 and 2, contained within Nilsson, M., Doctoral Thesis, *Restraint factors and partial coefficients for crack risk analyses of early age concrete structures,* Department of Civil and Mining Engineering, Division of Structural Engineering, Luleå University of Technology, Luleå, Sweden, 2003.

Palmer, D., *Concrete Mixes for General Purposes,* Cement and Concrete Association, London, 1977.

Parrott, L. A., A study of transitional thermal creep in hardened Portland Cement paste, *Magazine of Concrete Research,* 31(107), June 1979.

Reynolds, C. E and Steedman, J. C., *Reinforced Concrete Designer's Handbook,* Tenth edition, E and FN Spon, London, 1988.

Reynolds, C. E., Steedman, J. C. and Threlfall A. J., *Reynolds's Reinforced Concrete Designer's Handbook,* Eleventh edition, Taylor and Francis, Abingdon, 2008.

Sadgrove, B. M., *Water Retention Tests of Horizontal Joints in Thick Walled Reinforced Concrete Structures,* Cement and Concrete Association, London, 1974.

Scott, R., Forth, J. P., Mu, R., and Beeby, A. W., Test rig for shrinkage curvatures of reinforced concrete beams, *Strain,* 47, E551-54, June 2011.

Tammo, K. and Thelandersson, S., Crack behaviour near reinforcing bars in concrete structures, *ACI Structural Journal,* May–June 2009, 259–267.

Teychenne, D. C., Franklin, R. E. and Erntroy, H. C., *Design of Normal Concrete Mixes,* HMSO, London, 1975.

The Building Research Establishment, *Concrete in aggressive ground,* BRE Special Digest 1, 2005.

The Concrete Centre, *How to Design Concrete Structures Using Eurocode 2, 8 Briefing Notes,* 2005.

Vollum, R. L., Comparison of deflection calculations and span-to-depth ratios in BS8110 and Eurocode 2, *Magazine of Concrete Research,* 61, No. 6, 2009, pp. 465–476.

Wang, C.T., *Applied Elasticity,* McGraw Hill, New York, 1953.

Water Research Centre plc, *The Civil Engineering Specification for the Water Industry,* WRc, London, 2011.

Index